New Worlds in the Cosmos
The Discovery of Exoplanets

With the discovery in 1995 of the first planet orbiting another ordinary star, we know that planets are not unique to our own Solar System. For centuries, humanity has wondered whether we are alone in the Universe. We are now finally one step closer to knowing the answer. The quest for exoplanets is an exciting one, because it holds the possibility that one day we might find life elsewhere in the Universe, born in the light of another sun. Written from the perspective of a key player in the scientific adventure, this exciting account describes the development of the modern observing technique that has enabled astronomers to find so many planets orbiting around other stars. It reveals the wealth of new planets that have now been discovered outside our Solar System, and what this means in terms of finding other life in the Universe.

MICHEL MAYOR is Director of the Observatory of Geneva, Switzerland. In 1995, together with Didier Queloz, he discovered the first extrasolar planet (51 Peg b) around a main sequence star, and has discovered many more since. His work earned him the prestigious Balzan Prize 2000.

PIERRE-YVES FREI is a science journalist with the Swiss newspaper, Lausanne Hebdo. In 1998 he was awarded the Media Prize for science popularisation.

New Worlds
in the Cosmos

The Discovery of Exoplanets

MICHEL MAYOR

AND

PIERRE-YVES FREI

Translated by Boud Roukema

CAMBRIDGE
UNIVERSITY PRESS

PUBLISHED BY THE PRESS SYNDICATE OF THE UNIVERSITY OF CAMBRIDGE
The Pitt Building, Trumpington Street, Cambridge, United Kingdom

CAMBRIDGE UNIVERSITY PRESS
The Edinburgh Building, Cambridge CB2 2RU, UK
40 West 20th Street, New York, NY 10011–4211, USA
477 Williamstown Road, Port Melbourne, VIC 3207, Australia
Ruiz de Alarcón 13, 28014 Madrid, Spain
Dock House, The Waterfront, Cape Town 8001, South Africa

http://www.cambridge.org

First published 2003

Printed in the United Kingdom at the University Press, Cambridge

Typeface Trump Mediaeval 9.5/15 pt.   *System* LaTeX 2_ε [TB]

A catalogue record for this book is available from the British Library

Library of Congress Cataloguing in Publication data

Mayor, M. (Michel)
   [Nouveaux mondes du cosmos. English]
   New worlds in the cosmos: the discovery of exoplanets / Michel Mayor and
   Pierre-Yves Frei.
      p.   cm.
   Includes bibliographical references and index.
   ISBN 0 521 81207 0
   1. Extrasolar planets. I. Frei, Pierre-Yves. II. Title.
QB820.M3913 2003   523.24–dc21   2003041965

ISBN 0 521 81207 0 hardback

Contents

The plates will be found between pages 180 and 181

Figures

Preface

Do there exist many worlds, or is there only one?
That is one of the noblest and most exhilarating questions
in the study of nature.

Albert the Great (13th century AD)

It had to happen at some time or another that someone would look up at the sky and wonder about the nature of the stars. When did this first happen? Undoubtedly, long ago. The first explicit clues linked with astronomical activities date to several millenia before Christ. Just think of Stonehenge, the famous site in England, or of some of the ancient ruins inherited from the Sumerian and Babylonian civilisations.

It was with Greeks that astronomy started to distance itself from the influence of myth and religion. The sky, as well as the Earth, became an object of study, an object of observation, an object of science. Nature became less and less spiritual, and more and more material. However, the arrival of Greek thought did not stop speculation.

In the fourth century BC, the Greek philosopher Epicurus (341–270 BC) asked the fundamental and dizzying question: are we alone in the Universe? Nowadays we know that this question has a real scientific relevance. At the time, it was much less obvious. For the immense majority of Epicurus' contemporaries, at least for those interested in the question, the Universe was closed, bounded by a sphere on which the stars were fixed. But Epicurus did not see things this way at all. For him, the Universe was huge, so deep that it was impossible to determine its size. And in such vastness, he concluded that there had to exist infinitely many worlds, of which some certainly had to support life.

A lot of the history of astronomy tells the story of this quest for other worlds in some way or another. Over the centuries, from discovery to discovery, the cosmos has continued to swell and to be populated. Thanks to Copernicus, the Sun has taken the place it deserves, at the centre of the Solar System. With the arrival of the refracting telescope, Galileo and his successors discovered new planets. Saturn lost its claim to be the furthest planet from the Sun in the Solar System, handing the title over first to Uranus, then to Neptune and finally to Pluto. Uranus revealed itself first, then Neptune, then Pluto. It was also realised that the stars lie at incredibly great distances and that they are like other suns. There was no reason to believe that they had no planets. All that was needed was to prove that they did.

Today we have that proof. With the discovery in 1995 of the first exoplanet[1] around an ordinary star, 51 Pegasus, we know that the planetary phenomenon is not a privilege unique to the Solar System. So, Epicurus was right. In just seven years, over a hundred exoplanets have been discovered. Nearly all of them were detected indirectly, using the gravitational influence they exert on their associated central stars. This was the only way to be sure of their existence. And if the discoverers of the exoplanets are to claim any merit, then it could be that of having arrived at the time, at the turn of the twentieth century, when progress has provided a detection technique, that of the spectrography of radial velocities, which succeeded where others had previously failed, at times by a hair's breadth.

Already more than a hundred exoplanets have been discovered, and the astronomers' basket is still far from being filled. Massive resources have now been committed in order to push this number to several hundred, and undoubtedly in the near future to several thousand. This is because we need to have a large number of these exoplanets in order to better understand the conditions that led to their formation. True, a beautiful theory of planetary formation had already been constructed, based on the study of all sorts of properties

---

[1] We have got used to calling planets outside of the Solar System 'exoplanets'.

of the Solar System. But unfortunately or fortunately, according to one's view, the exoplanets discovered up to now aren't at all like those of our solar system. When they're not virtually stuck to their central star, they have particularly eccentric orbits. All the data point to a need to reformulate most of the theory of planet formation. A challenging perspective.

But what is even more exciting is the very realistic prospect of discovering life elsewhere than on Earth. Science fiction has already given us a tempting foretaste of this. It underlines our overwhelming desire to know if we are alone in the Universe. At the moment, we can't say anything definitive. But it is hard to believe that among the billions and billions of stars in our Galaxy, the Milky Way, and the billions of galaxies that inhabit the Universe, the Sun is the only one accompanied by a living planet. It's likely that life has conquered other planets, even if only in a primitive, unicellular form. In the Solar System, life could have chosen other playgrounds: there's Mars, a quasi-twin sister of the Earth, but also Europa, one of the four big moons of Jupiter, which could hide marvels beneath its frozen surface. Several space missions are planned that will look for signs of past or present extraterrestrial life.

Nevertheless, one thing is sure: the Solar System does not harbour another blue planet. To find a cousin of the Earth, it will be necessary to go further, maybe even much further. The closest star to the Sun is 4.2 light-years away, a considerable distance, especially when you are trying to observe a planet which produces no light of its own. So, is it mission impossible? Not if one believes the astronomers, who have already thought of techniques that will make it possible to 'see' these extrasolar earths. Some projects, more ambitious yet, but still scientifically sound, plan to photograph those same earths, at least if they're not too far away. In which case, we will see them in their true colours, in blue, white and brown. Will we see the first extrasolar portrait before 2050?

If, by good fortune, such a pearl of life is found, it is certain that radio astronomers, some of whom have been scrutinising the sky since the 1950s in search of an extraterrestrial signal, will direct their antennae towards this new world. Maybe, humanity will then finally learn that we are not alone in the Universe.

The aim of this book is to provide its readers with a glimpse of the quest for exoplanets, to show them that even though this discovery constitutes an important event, it's only a link in the tremendous chain of knowledge and questions which leads to what will possibly be one of the greatest moments in the history of Mankind.

Acknowledgements

I should emphasise here that the discovery of the companion of 51 Pegasus was not the work of just one or even of just two researchers. Astronomy today requires ever more powerful and sophisticated instruments, the development of which cannot occur other than as the result of close teamwork.

The spectrographs that we used in order to obtain our harvest of planets would not have seen the light of day had it not been for the invaluable help of numerous specialists from the Observatories of Geneva and Haute-Provence, in domains as diverse as optics, mechanics, electronics and, of course, computer hardware and software. I'm particularly grateful to my friend André Baranne, the 'magician' to whom we owe the extraordinary quality of the optics of the Élodie spectrograph, the true cornerstone of the experiment. A very big thank you also goes to Didier Queloz not only for his very pleasant company during our numerous – and in certain cases historic – observing campaigns, but also for his skills in perfecting the software that presently runs Élodie and without which our instrument would never have reached its full potential.

To go from a simple idea to the completion of such complex and precise machines took many years of effort, years during which at times there were moments of doubt. Luckily, we have been able to count on the support, the help, the faith and the enthusiasm of many people. They should all be thanked for having contributed to this beautiful adventure: Philippe Véron, Dominique Kohler, Alain Vin, Georges Adrianzyk, Jean-Pierre Meunier, Gérard Knispel, Luc Weber, Daniel Lacroix, not to mention all our other friends at the Observatories of Geneva and Haute-Provence.

The discovery of the exoplanets has given rise to an upsurge of new questions on the origin and formation of the stars and planets in the Universe. There are many challenges, and in order to meet them, our team has expanded considerably during the last five years. The present team is a gathering of people whose personal qualities match their scientific expertise. I would like to take the opportunity here to thank them all and confirm my appreciation of their qualities: Jean-Luc Beuzit, Dominique Naef, Francesco Pepe, Christian Perrier, Nunõ Santos, Jean-Pierre Sivan and Stéphane Udry.

I would also like to remind the reader that the discovery of the planet orbiting 51 Pegasus and its sister planets will not be enough in itself to bring all of the mystery of the exoplanets to light. It's only thanks to the help from many scientific disciplines that all these questions will one day be clarified. During the writing of this book, we have been able to take advantage of the expertise and observations of many specialists, who have played their part with patience and kindness. Thank you all: Jean-Philippe Beaulieu, Willy Benz, Xavier Delfosse, George Gatewood, Jean-François Lestrade, Andrew Lyne and Alexander Wolszczan.

Finally, this book is the result of a meeting between Pierre-Yves Frei, the science journalist, and myself, the astronomer. We together decided to set forth on this project. Hours and hours of interviews later, this stimulating collaboration has finally borne fruit. We hope that this work will be accessible to a wide readership. In any case, that's what we wanted. Happy reading!

1   The quest begins

The discovery of the exoplanets is undoubtedly a technological feat. But without exceptional people, there would not be any technological feats. So it's equally – and maybe even primarily – a human adventure, a personal experience, and, for this reason, is better narrated in the first person singular, in the voice of Michel Mayor. This will essentially be the case in the first chapter, but also every now and then in later chapters (especially Chapters 6, 7 and 8). These uses of 'I' are like memories which suddenly bubble up to the surface during particular scenes. And during the numerous interviews that took place between the scientist and the journalist, there were many of these bubblings. So, henceforth, 'I' and Michel Mayor will be indistinguishable.

Having made this comment, all that remains to be done is to set the scene by beginning with a trip through time and space. It's October 1995, in Florence, the seductive Tuscan town where art and science live hand in hand in idyllic happiness. It's there that it all starts, where the discovery would see the light of day.

For Didier Queloz, my young collaborator, it's his first trip ever to Italy. Given the (happy) circumstances which brought us here, he decided to celebrate the event, sharing a room in a beautiful hotel, combining luxury and quaint charm, with his wife, Valérie. Each evening, a housekeeper elegantly prepares their bed and brings each of them a pair of slippers. My wife, Françoise, and I have chosen a nice, small hotel which I liked the last time I came here.

The conference which we are participating in should last from 2 to 8 October 1995, and my presentation is planned for the 5th. This week of science is devoted to cold stars. This is what, in our specialist jargon, we call stars that are the least hot. By this yardstick, our Sun, with its average surface temperature of 5800 °C, is a member of the cold star family. It's far too cold to compete with stars with temperatures above 20 000 °C.

Paradoxically, Didier and I are not here to talk about stars, instead we want to talk about planets. Or rather about *a* planet, the first one ever discovered outside of the Solar System around an ordinary star. We discovered it a few months ago around 51 Pegasus (also called 51 Peg), a star more or less like the Sun, located in the northern constellation called Pegasus.

Officially, we're bound to secrecy until the public release of our article in the British magazine *Nature* in a few weeks. But the editors finally agreed that we could give a lecture during this conference. In any case, the rumour according to which a team from the Observatory of Geneva had possibly discovered an exoplanet (a planet located outside of the Solar System) has already been circulating for some time among astronomers.

I arrived in Florence, and the hotel receptionist gave me the key to our room, together with an impressive pile of faxes. It was clear that the rumour had spread beyond the private circle of astronomers. It had reached the general public. Newspapers from around the world were asking me and even begging me to call them back as quickly as possible to agree on an interview regarding the first exoplanet around an ordinary star. Unfortunately, there was nothing that I could do. The *Nature* team were unequivocal. They agreed that we could talk to our colleagues, but they forbade us from giving the merest hint of an interview before publication of our article.

This half-solution yielded an ironical result. After my presentation, I was obliged to refuse to answer all the questions fired at me by the press, radio and TV journalists, who, instead, turned to my colleagues for comments. Everybody's talking about our discovery. Except for us! Didier and I watched this ballet without bitterness. We'd been overtaken by the pace of events. Our nights of work at the Observatoire de Haute-Provence (OHP) were certainly not enough to prepare us for all this fuss.

STARS FIRST

I was not born – in 1942, by the way – as a planet hunter. In fact, I got there rather late, many years after the great American experts Geoffrey Marcy and Paul Butler, or the Canadians Gordon Walker and Bruce Campbell. However, as far back as I can remember, science has always attracted me. Whether it's physics, chemistry or biology, I've always enjoyed exploring the secrets of Nature. Where does this curiosity of mine come from? I haven't got the foggiest idea. My parents were not scientists. Maybe I owe some of my interest to Edmond Altherr, an extraordinary man who was responsible for teaching the sciences in my college in Aigle, in the Vaudois region. He had studied biology and wrote a thesis on nematodes, those microscopic worms much utilised by geneticists today. Hundreds of Aigle children are indebted to this man not only for his knowledge of nematodes but also for his enthusiasm, pedagogical sense and most especially for his lectures covering the full glory of Nature, the flowers, trees and animals.

After completing my high school final exams, known as the 'maturity' in Switzerland, or the 'baccalauréat' in France, I found myself at the threshold of starting university. I hesitated between studying maths and physics, and finally chose physics. I hesitated again after obtaining my first degree in 1966. Two thesis subjects were proposed to me. The first was to do with solid state physics, the second was in astrophysics. The decision was made during drinks with a friend from my graduation class, who was, I think, as undecided as I was. Finally, he chose solid state physics while I chose the road to the stars. Maybe because I liked watching the sky and the flickering of stars when I was a young scout and we spent the night outdoors.

I joined the staff of the Observatory of Geneva, and I got stuck into stellar dynamics, a subject bubbling with interest. At the time, astronomers were especially excited about the spiral arms of certain galaxies such as our own, the Milky Way. How, they asked themselves, is it possible that such structures, measuring up to 100 000 light-years in size, could not only be created, but also remain in existence for hundreds of millions of years? A galaxy doesn't rotate as a solid body. Its main elements, stars, don't support each other, even if they do influence one another. For the stars at the edge of the galaxy to remain perfectly synchronised with the stars at the galactic centre, they would have to accelerate at speeds which are simply unimaginable. An impossible feat. So, the heart of the galaxy turns faster than its edges, in a grand ballet that specialists call 'differential rotation'.

This particular aspect of galactic dynamics provides a good explanation of astronomers' observations, but it fails to enlighten us as to why certain galaxies, such as the Milky Way, have huge spiral arms, usually two, sometimes four.

This puzzle was solved by two American theorists, Frank Shu and C. C. Lin. They proposed, in a paper that became very famous, the existence of density waves in a galaxy, which were generated by the mass in the galaxy. These waves were proposed to propagate through the carpet of stars with a constant angular speed: in other words, like a windscreen wiper that sweeps the galaxy, slowly, but uniformly. It is these waves that create the spiral arms. Not only do the waves create the spiral arms, but inside of the arms, they initiate the gravitational collapse of numerous clouds of interstellar gas into tight, dense objects, which light up and burst forth to the eye as myriads of bright, blue, young stars.

The pioneering work by Shu and Lin trailed in its wake an explosion of related research. My thesis was part of this revolution. I had to study one particular aspect of their theory. The two Americans had predicted that stars that fall into the field of one of these gravity waves must initially slow down, but then accelerate again once they are freed from the grasp of the wave. Are these changes measurable on small scales? In other words, among stars in the solar neighbourhood, is it possible to discover differences in speed that reveal the discreet, but significant influence of these cosmic waves? These were the questions that I had to tackle.

There was not much choice about what had to be done. My absolute priority was to gather data on stellar speeds in the solar neighbourhood. In order to avoid long and boring observations, I delved into stellar catalogues, patiently compiled by generations of astronomers, which gathered together thousands of stars classified according to their properties. It was a painstaking task and I found that samples were too small, measurements were not precise enough, nothing really satisfied me. The technology of the time had its limits, and it's these that pushed the theorist that I am to think about instruments and the way to improve them. But I lacked the know-how. Luckily, a meeting with a man and his machine allowed me to progress from thought to action.

In the summer of 1970 I went to Cambridge to participate in a conference on one of the most intractable problems of stellar dynamics, that of N-body systems, which has to do with gravitational interactions between many objects (N bodies). This is a research field that concerns galaxies as much as stars.

George Contopoulos, an eminent theorist from the University of Athens, was also at the conference. I was itching to ask him some questions related to my thesis. When I asked him to spare me a few minutes, he just gave me a blank look. Clearly, I was not the only one wanting his attention. His diary was overflowing with appointments and I was only a twenty-eight-year-old postgraduate student. Despite everything, he agreed to listen to me while we visited the domes of Cambridge Observatory, which, I think, don't particularly excite him. It's just that Contopoulos is a pure-blooded theorist, one of those who don't pay much attention to the world of instruments.

I was so busy listening to Contopoulos that I barely glanced at the telescopes that were being presented to us until we entered the umpteenth dome, where a man of some thirty-five years was going about his business. Roger Griffin is one of those scientists fashioned by legend, who cultivate solitude and discretion with the same intensity. He's a perfect example of man–science symbiosis, an observational fanatic. His house is less than a kilometre from Cambridge Observatory. Each night, he pokes his nose outside, looks at the sky, and if the night is clear, jumps on his bicycle and rides to the Observatory. It's easier to understand his haste when you know just how little suited the English climate, often grey and rainy, is to the study of the stars.

A REVOLUTIONARY INSTRUMENT

At the time, Griffin was working on a new spectrograph, which was inspired by the work carried out fifteen years earlier by one of his compatriots, Peter Felgett. The radial velocity spectrograph, as it is called, measures one of the components of a star's velocity. Projected on the celestial sphere, stars can move either to the left or right, or up or down. And if they move fast enough, their movement becomes visible over the years. Also, the same stars can either approach or recede from the Sun along our line of sight (the radial viewpoint). But we cannot observe this latter movement by eye. It's here that the spectrograph comes into play, due to its ability to decode all sorts of light, in particular that of stars. By separating out the light rays, according to the principle of the prism, the spectrograph shows whether a bright source is approaching or receding, and at what speed it is moving.

The apparatus that Griffin was using to study radial velocities was a huge improvement relative to everything else that was available at the time. In relative efficiency, his equipment was a thousand times better than the best instrument elsewhere. I understood in that instant, while listening to the words of my British colleague, that his ingenuity had sky-rocketted us from the age of the wooden wheel to that of the Formula 1 tyre! The power of this new spectrograph was exactly what was needed to revolutionise stellar dynamics, to create star catalogues of unheard-of precision, and finally to look for much less massive binary stars than was possible in the past.

A binary star is a system of two stars, close enough together that they attract one another. Sometimes, the two components are of equal mass and luminosity. Sometimes one of them is so faint, drowned by the light of its neighbour to such an extent, that it's invisible to us. In that case, how can we know whether a star is solitary or not? There are various techniques. One of them is based on the influence that every body exerts on every other body due to the force of gravity. This is a universal law. Just as the Sun attracts the Earth, the Earth attracts the Sun, except that it does it in proportion to its tiny mass.

This reciprocal attraction is particularly valuable because it's this that allows us to see the invisible, to detect a star too faint to see, by recording the perturbing movements that it causes on the main star. Under the gravitational influence of its companion, the main star moves over a small circular path, which is revealed by changes in the light that it sends us. The lighter the companion and the heavier the main star, the more subtle these changes are. Thus, the greater the sensitivity and precision of a spectrograph, the better it can detect the slight perturbations of the stars whose light it analyses.

It was by increasing precision that, many years later, we would succeed in detecting some of the first exoplanets. Several years passed, inspired by these new, useful techniques. At the time, even though it constituted a real technological breakthrough, Roger Griffin's instrument was only a prototype: with rudimentary electronics, cogged wheels and lamps. That spectrograph deserves to be displayed in a scientific museum as an example of high quality do-it-yourself. You have to realise that given only a few thousand francs of funding, my British colleague was condemned to a pretty heterogeneous result. Luckily for him, his genius inspired him well. As an excuse for a cooling system, he put end-to-end an old refrigerator, a fan and a tray of silica-gel which he used to prevent the humidity in the air from condensing and blinding the spectrograph. As for thermal insulation of the mirror, he tied a down jacket to the frame with a few strings. However, despite this unbelievable construction, the instrument worked marvellously.

After Griffin's spiel, I continued to fire questions at Contopoulos during the rest of the Observatory visit. I didn't see Griffin again during my entire stay, but back in Geneva, I talked of nothing but him and his spectrograph. I had to persuade my Genevan colleagues to try to build a spectrograph ourselves. The answer given to me by Marcel Golay, the director of the Observatory of Geneva at the time, was 'Well, if you want to do it, then do it!' He was challenging me to put my words into action, undoubtedly to see how determined I really was.

Before thinking of how to finance the spectrograph, I had to start with some theoretical calculations. Some of my colleagues gave me some heavy duty help in making the first numerical simulations. The optical system of the telescope was a particular source of worry. I understood nothing, or next to nothing, of the subject. I really tried to read several specialised papers, but in vain. Optics was another world, the inhabitants of which were scarce. At the time, I only knew one, someone famous: André Baranne, from the Observatoire de Marseille.

Although not expecting his active participation – he is in great demand – I hoped to get some advice from him. I telephoned him to fix an appointment. I hoped to take advantage of a week of work at the Observatoire de Haute-Provence, in May 1971, in order to go to Marseille and meet him. André immediately agreed to my visit. He's not the sort who wastes time in procrastination. He's very direct, to the point of talking quite bluntly, with everything implied by this when it comes to making friends or enemies. Born in 1933 in Bagnères-de-Bigorre, a region that breeds many rugby players, André is of a thickset build, which perfectly matches his straight talking character. He is someone who enjoys life, taking as much pleasure from eating as laughing. My three children have greatly enjoyed his tall tales, told in his magnificent Gascon accent.

I couldn't have been more surprised when, a quarter of an hour after the beginning of our interview, André Baranne burst out with: 'OK, I'll handle the optics!' Not only did he not send me back to square one, but he wanted to be part of the adventure. I couldn't contain my excitement. With astonishing speed, he spotted all the difficulties, all the points with which he would have to deal. He was already outlining solutions. Without a doubt, I had before me a worthy representative of the École supérieure d'optique in Paris, whose reputation goes well beyond the French border.

André immediately came up with a trick of optics that would improve the performance of our spectrograph. Without going into details, it's a certain way of separating the light arriving from stars in order to give a two-dimensional spectrum, so that much more precise measurements become possible. At the time, nobody believed this technique would work. Two studies concluded that it was not feasible. Ironically, one of them was written by Roger Griffin himself. I also remember that Jim Brault, an American optical astronomer visiting the Observatory of Geneva, had told us how pessimistic he was of our chances of success. But it would have required a lot more to shake the faith of André. He was convinced of the feasibility of the technique. He was right.

I put the finishing touches to my thesis in the summer of 1971. The day after submitting the thesis, I jumped on a plane heading to London. Roger Griffin, the astronomer magician, had agreed to act as my host for a few weeks, giving me time to understand how the spectrograph worked. This was a big task, but the friendliness of my British colleague lightened the workload. I returned from Cambridge with my head full of plans and projects, and back in Geneva, I applied for a grant from the Fonds national suisse de la recherche scientifique (FNRS). In 1972, I obtained a hundred and fifty thousand Swiss francs [translator: roughly a hundred thousand euros, ignoring inflation] for a period of two years.

Roger Griffin's instrument had certainly made a deep impression on me, but I was also alarmed by the large number of hand calculations required. Much of the data reduction was done by ruler, pen and paper. Griffin could do this with uncanny ease, but I personally didn't have the stomach for it. I preferred to do all calculations on a computer and let a young doctoral student in physics specialising in electronics, Jean-Luc Poncet, do the work of programming the machine.

It took us five years to design and construct our two spectrographs, one for each terrestrial hemisphere. The first was installed at the Observatoire de Haute-Provence (for the North) and the second in the Chilean observatory at La Silla (for the South). Five years can seem like a long time. However, in the early 1970s, computers had neither the speed nor the power of those of today. It turned out to be painfully difficult – apart from opting for a perfected machine, which was beyond our means – to get them to carry out several tasks simultaneously. But multi-tasking was something we direly needed. Jean-Luc Poncet had no choice other than to completely rewrite the operating system. We wanted our computer to be a sort of computerised Shiva, able to seize everything that we prepared for it with its multiple arms.

For a long time the multi-tasking programme remained too big for the memory of our machine. Eight kilobytes was the entire capacity that we had at the time, compared with the basic office computer of today which can easily store a million times more. Two months before the date at which it was planned to put Coravel (CORrelation Velocity), the spectrograph destined for the OHP, into service, Jean-Luc, after having tried all the tricks he could think of, gave up. Nothing helped. The programme refused to be adapted to the tiny memory. Luckily, we found a second-hand memory extension card for the modest sum of twenty thousand francs. Well, that was the dawn of the computer age!

The first experiments using Coravel started in April 1977, after it had been attached to the 1-metre telescope at the OHP. As the designer of the instrument, I had to take on certain additional tasks. Even though, thanks to computerisation, the spectrograph gave almost instantaneous measurements, these still required further refinement before being catalogued. It was the work of a beast of burden which had to be added to my other research. In essence, the latter consisted of looking for all the ways in which Coravel could be used. It turned out, for example, that Coravel is perfect for measuring the radial velocity of stars, and for measuring their rotation and the elementary composition of their atmospheres.

GRADUALLY APPROACHING THE PLANETS

Five years later, I decided I needed reinforcements. There was a young Frenchman, Antoine Duquennoy, who had been at the Observatory of Geneva for several months. He was an astronomer and a signal analysis expert who had obtained permission to carry out his military conscription duties with us. He was a lad from Amiens, introverted, very discreet, highly intelligent and gifted with a rare capacity for hard work. I became an employee of the University, as a substitute lecturer, just as he was about to finish his period of conscription. Using the money left from the FNRS, I hired Antoine. We were to work together for more than twelve years. It was a very fertile period. Antoine's death in a car accident was a tragic event. He died six months before the discovery of 51 Pegasus' companion.

Right from the beginning of our collaboration, I threw Antoine into the field of double stars. These turned out, by the way, to be his main thesis subject. The work that he carried out remains to this day the most beautiful and the most precious ever carried out with the Coravel spectrograph. I can say without the least hesitation that it's thanks to this that a few years later we became planet hunters, even if, at the beginning of the 1980s, we weren't yet thinking of looking for them.

The Sun is one star among billions of others. Its isolation distinguishes it from most other stars of its class. The vast majority of stars, about two thirds of them, live in a partnership or in three-, four- or five-member families bound together by their mutual gravitational attraction. At the time, we had very little data on binary systems. We knew they existed, but, due to the lack of an appropriate instrument, there had been no systematic classification of their properties (distributions of orbital period, mass, orbit shapes, etc.). These are all data that have to be gathered if we want to understand better where stars come from. So, we had to decide on a uniform sample. We chose to concentrate on 164 binary systems of which the main star was a G dwarf, similar to our Sun, and which are within at most 72 light-years from the Sun.

I had already started this classification when Coravel became operational, but it was Antoine who continued where I left off and did the bulk of the work. This task kept us busy right up to 1991 when it was published. During the whole period, we never stopped refining our instruments and gaining in experience. Our hunger sharpened. As our habit of probing for stellar companions developed, we looked for the smallest we could find, those that flirted with the limits of our instrument.

At the time, astronomers were looking for a mysterious type of star, called a 'brown dwarf', a star whose existence is only theoretical, a sort of missing link, having a mass which lies between that of the smallest stars and that of the biggest planets. Brown dwarfs are failed stars, unable to maintain nuclear reactions over long time scales, that end up being extinguished, condemned to live a hidden life.

At the end of the 1980s, many teams were hunting for brown dwarfs. Some were looking for them as companions of normal stars or of white dwarfs, stars which are as massive as the Sun but have only the diameter of the Earth. Others preferred to hunt free brown dwarfs, either by scrutinising young star clusters like the Pleiades, or by passing a fine tooth comb through the halo of the Galaxy. Announcements of discoveries came one after another. Most were dismissed after further analyses. The others remained unconfirmed, because it's so easy to mistake one of the smallest and faintest normal stars, which we call red dwarfs, for a brown dwarf.

This was when David Latham, from the Harvard-Smithsonian Center for Astrophysics, contacted us. Like us, he was using a spectrograph in order to detect stellar companions indirectly. He thought he had noticed a suspicious oscillation in the star HD 114762 which could be due to a brown dwarf or to a giant planet with an 84-day orbital period. It happened that HD 114762 was in our sample. We had looked at it several times. Ironically, we were using it to calibrate our measurements. According to the catalogues, it was an extremely stable star. In fact, it's not quite that stable after all. It's simply that the earlier instruments were not powerful enough to tell. Ours were. In 1985, we thought we noticed an anomaly, but it was not big enough for us to take it seriously. When David Latham brought our attention back to this star, all that we needed to do was to check through our data in order to confirm his discovery.

Our contribution was recognised when we coauthored the article that appeared in *Nature* in 1989, entitled 'The unseen companion of HD114762 – A probable brown dwarf', a cautious title, for good reason. Calculations showed that the mass of our invisible object was at least ten times that of Jupiter. And since our detection method can only give a minimum estimate, it was therefore quite possible that the object was a brown dwarf rather than a planet. In any case, for the first time ever we had identified a low mass companion.

Our scientific reputations in the field of low mass stars gained a certain amount of credit thanks to our being part of this discovery. At the time, we were already thinking of exoplanets. In 1990, we participated in a conference organised at Val-Cenis, in Savoie, devoted to bioastronomy, that is, to research into extraterrestrial life. Several heavyweights were there: the Canadian Bruce Campbell, the American David Latham, the Frenchman François Raulin, a great expert in prebiotic chemistry (the chemistry leading to the appearance of life), Jill Tarter, the famous radio-astronomer who inspired Jodie Foster in her role in the film *Contact*. If my memory's not playing tricks on me, Frank Drake, the author of the famous equation for the probability of finding intelligent life in the Universe, was also at the conference. For most of us, it was the first time that we had met. It was like an initiation ceremony, like being dubbed. We were admitted to an order of knights sworn to the quest for an ephemeral, astronomical Grail. We openly became planet hunters.

Our work started bearing fruit and we were able to plan ahead. Also in 1990, Philippe Véron, the director of the Observatoire de Haute-Provence, told us about his intention to complete the instrument suite so that observations would be possible even during full moon nights. André Baranne and I jumped at the chance. It was time to build the successor to our first spectrograph, Coravel, using everything that modern technology put at our disposal: digital videocameras, optical fibres, etc. The new instrument was to be attached to the largest of the OHP's telescopes, the 1.93-metre. Our future spectrographs, one planned for the OHP and the other for La Silla (Chile), were to be called Élodie and Coralie respectively.

Once again, computers were a bottleneck. It was up to a young PhD student from the astronomy institute to tackle this. Didier Queloz is a Genevan, a lean, lanky fellow, all smiles and enthusiasm, with a frankness matched by his spontaneous and generous sense of humour. I set him to work on something which had little to do with classical spectroscopy. At the time, we had to start thinking about the future of our technique. The arrival of large instruments, such as the Very Large Telescope (VLT), which has just been built in Chile at the Paranal site, was likely to cause upheavals in our profession. It would be best to anticipate these rather than be caught out by them. Didier busied himself with this strategic planning task during his diploma year. It seemed natural to me that he should continue it for his thesis.

The development of the computer programme for the Élodie spectrograph was not meant to be anything more for him than a job on the side, just part-time work taking up a little of the time spent on his doctorate. I expected that the work would occupy him for, at most, a year and a half or two years. In fact, it took him more than three years to complete the task. It's not that Didier wasn't up to it. On the contrary, it's largely thanks to him that the precision of Élodie was three times better than it was planned. I had simply underestimated the workload required in developing the programme. In the end, the programme constituted a major fraction of Didier's thesis.

Élodie was installed in June 1993, but it only became usable a year later. This was due to an optics problem. Like any newborn, our spectrograph demanded, at least to start with, constant attention. We felt a bit like fairies leaning over the cot, filled with hopes, impatient to know whether or not our infant would turn out to be as talented as we hoped.

During this period, the quest for the exoplanets, while not suffering any setbacks, did have at least one disappointment. In 1994, two independent groups, the American pair Geoffrey Marcy and Paul Butler, and the Canadian pair Gordon Walker and Bruce Campbell, who have very good spectrographs, published their intermediate results. Neither group had found any exoplanets. This was hardly encouraging. However, our method and our approach were different from theirs, so we felt that they were worth trying.

Due to the sensitivity limits of their spectrographs, planet hunters of the time could only hope to detect massive objects, giant planets at least as big as Jupiter. But Jupiter takes just under twelve years to complete an orbit around the Sun. If all extrasolar, gaseous giants had similar orbits to Jupiter, then planet hunters would have to follow and measure stars over many years in order to have sufficient data to identify a companion indirectly. This is why our American and Canadian colleagues examined only about twenty stars.

Our own preference was to go for big stellar samples. This is because we were primarily looking not for planets, but for low mass stellar objects such as brown dwarfs. At the time, we already knew that it seemed that the latter did not like coupling to solar type stars. But we needed more observations in order to hone our statistics. Luckily, theory does not forbid brown dwarfs from gravitating, in a binary system, very close to their main star, and to make a complete turn in just a few days. Two stars being close implies a strong gravitational influence between them. This is an ideal situation for using the radial velocity method which can detect the numerous, fast oscillations of such cosmic couples and so reconstruct, after just a few days of observation, the characteristic orbital period. Since these sorts of data can be collected 'quickly and easily', we had no reason, unlike the planet hunters, to limit the size of our star sample. And if by chance an exoplanet should happen to be caught in our net, so much the better. Luck turned out to favour us.

51 PEG'S COMPANION

In November 1994, Didier Queloz was alone handling Élodie, our new spectrograph. I was at the University of Hawaii. My visit was to last six months, during which I hoped to attend seminars about the new giant telescope on the island, the Keck, as well carrying out observations with my French colleague Christian Perrier on the 3.60-metre Canada–France–Hawaii telescope (CFHT). During this period, my collaboration with Didier consisted of only a few telephone calls and email exchanges.

The planetary system of 51 Pegasus

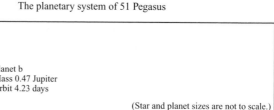

51 Peg

Planet b
Mass 0.47 Jupiter
Orbit 4.23 days

(Star and planet sizes are not to scale.)

58 million km

Sun

Mercury
Mass 0.055 Earth
Orbit 87.9 days

All graphics are by Pierre-Yves Frei.

At the time, we had a week of observation available every two months. The key work consisted of calibrating our new instrument in order to eliminate all sorts of artefacts that could possibly contaminate our measurements. Didier regularly checked out ten reference stars. This is when he thought he detected an oscillation of one of these. The movement was clear and rapid, which was surprising for a supposedly stable star. Of course, it could have been an instrumental problem. But if that were the case, then why did other stars not vary in the same way?

Thus, 51 Peg ceased to be, for us, a plain, standard star. It became an important target, the centre of our curiosity, our inseparable companion for our nights of observations. It was crucial to keep a cool head. Even after a barrage of tests that failed to find any instrumental error, other potential sources of error remained ten-a-penny. 51 Peg oscillated, that was for sure. But maybe it was just internal stellar pulsations, or an effect due to sunspots or atmospheric instabilities.

Didier continued to make observations. From the other side of the planet, I followed his progress. Measurements accumulated, and became more and more precise. Alone under the dome in

Haute-Provence, in the middle of clear winter nights, Didier went through the whole gamut of emotions. If it were really a planet, then what a fantastic adventure this would be for such a young researcher! At the end of February, we had eighteen measurements accumulated over a period of 150 days. Enough of a harvest to make a few estimates. If there were a stellar companion, then it would have to have at least half the mass of Jupiter, while the orbital period would be 4.2 days. This would mean it was a planet that goes around its star in just over 4 days! This seemed incredible! It implied that the companion was twenty times closer to its star than the Earth is to the Sun. No theory had predicted such a situation.

We calculated an ephemeris, i.e. a model able to predict the planet's movements, supposing that it really was a planet, using the data we already had. If our future measurements confirmed the ephemeris, we would have hit the jackpot. Normally, a month would have been long enough to make the necessary measurement but this does not take into account the restrictions due to celestial mechanics. At European latitudes, in March, the Pegasus constellation, in which 51 Peg is to be found, hardly rises at all in the night sky. Lying just above the horizon, it becomes difficult to observe. Our Holy Grail disappeared in the spring sky without giving us the time to compare our model with real data. We had to wait until the summer to find out more.

We didn't take it easy during those few months without Pegasus. We couldn't keep this mysterious object out of our minds. Its strange properties, so unexpected, so clearly absurd, obliged us to delve into specialist literature with which we were not terribly familiar. We were transformed from nocturnal observers into daytime bookworms. Once I returned from Hawaii in April 1995, Didier and I got back to work with glee. It was a sort of *ménage à trois*: two men and a star.

We did our utmost again and again to try to turn the planet orbiting 51 Peg into a mirage. We had to search for all the possible sources of error. There had already been so many false discoveries and premature announcements in the quest for exoplanets. These errors, while being an unavoidable component of scientific research, had weakened the credibility of our science. In fact, my American colleague Geoffrey Marcy once told me about his reaction on learning about the discovery of a planet around 51 Peg (he wasn't present at the Florence conference), 'Once again a hope which I'll be obliged to dash.' Though he's very friendly and kind, Marcy is nicknamed 'Dr Death'. It's true that his very rigorous counter-analyses have 'killed' many an exoplanet candidate.

Scientific knowledge has attained such a high level today that it has become incredibly piecemeal. So, to resolve certain mysteries related to 51 Peg, we urgently needed external advice. But at the same time, we wanted to keep our discovery absolutely secret until we had sufficient information. Which is why we resorted to verbal acrobatics. I was still at the University of Hawaii when I received a message from Didier Queloz confirming the incredible orbital period of about 4 days for 51 Peg's companion. I decided to ask Ted Simon, a cold star specialist, for advice. From the way he looked at me I understood that he was wondering whether I'd lost my marbles. Undoubtedly he was asking himself why anyone would waste time with these sorts of absurd hypotheses. For anyone unaware of our data, such scepticism was well justified.

Finally July arrived and with it a new 51 Peg observing campaign. It was 6 July 1995. Our families came with us to Provence, taking advantage of the school holidays. We wanted them with us to celebrate the event, if there were to be an event. Our ephemeris predicted a precise value of the radial velocity (see Glossary) for the date of 6 July 1995. All that we had to do was to wait for nightfall, enter the dome, point the telescope at 51 Peg, measure it and compare the result with the predicted value.

Everything happened like in a dream. In just a few moments, the computer gave the verdict. It was exactly what we hoped. It was a spiritual moment. Our planet exists; we were now virtually sure of this. Out there, 42 light-years away, a gaseous giant of at least half the mass of Jupiter was orbiting a star similar to the Sun, but a bit older, in just over 4 days. With an infinitesimal blink of an eye, it had revealed its presence to us and promised a marvellous scientific adventure. That evening, we honoured the breakthrough with sparkling wine and a delicious cake bought in a Manosque cake shop.

It was a twist of fate that this historical observing campaign occurred at exactly the same time as a conference on the search for exoplanets organised by the OHP. This was a very important meeting. Its conclusions would be used by French scientific committees who were to decide whether or not to invest in programmes for looking for low mass stellar companions. So while at night we flirted with our new planet, during the day we played complete innocents, participating in the conference, presenting the general approach of our own research. Of course, I was dying to reveal all to my French colleagues who were discussing the relevance of investing in such an uncertain quest, but we had to keep mum. Not only is it contrary to the norm to announce a discovery without having previously submitted it to peer review, but there is an additional danger when talking of an unofficial discovery because there's a real risk of being overtaken at the finishing line.

We started to write our article destined for *Nature* in the middle of the summer of 1995. We worried incessantly about having missed some important detail. At the time, we had the pleasure of a three-month visit by Willy Benz, one of my ex-doctoral students, who had since become a professor at Tucson, Arizona. A specialist in astrophysics, and also in planetology, he made a perfect proofreader. He looked over the first paragraph and smiled, undoubtedly amused by the absurd period of 4.2 days. But he promised, nevertheless, to read our article more carefully that very evening. The following day, seeing his expression, I realised that our arguments had convinced him. He even asked me why we had waited so long before publishing. Our answer was prudence.

There was still something nagging us. Did our planet have a reasonable life expectancy or was it as fragile as tissue paper? Two dangers lay in wait for our planet which dared, like Icarus, to approach its sun so closely, only 7.5 million kilometres away. On one hand, it risked being pulled to bits by the tidal effects due to the star. On the other, it risked being vaporised over time by the intense heat surrounding it (we estimated that its surface temperature would be about 1300 °C). To discover more about this, Willy decided to contact his Tucson colleague Adam Burrows, the leading expert on cold stars.

The problem was presented to him as a purely speculative exercise. But Adam Burrows, who had the whiff of the scent, asked my ex-student if his Swiss friends had by any chance detected something with their spectrograph. Willy ducked the question. The American, playing fair, didn't press the point. He fed the question to the number crunching power of his computers. His simulation software determines the minimum distance from its star that a planet, a gaseous giant, for example, can orbit and remain stable. Two days later, Burrows delivered his verdict. His programme showed a divergence below 4.5 million kilometres. Since 51 Peg's planet is at 7.5 million kilometres, it could be considered to be stable in the long term, theoretically, at least.

On 25 August 1995, we submitted our paper to the magazine *Nature*. Simultaneously, we thought about when it would be best to announce our discovery. The Florence conference which was to take place in October suited us particularly well. The intervening time was enough to allow time for the three experts who would review the paper to give their verdict on the content of our research. Also, since the conference only took place every second year, it was very much the 'in thing' and by a stroke of luck more than three hundred specialists were expected.

As it was too late to register as a speaker, I asked to have some time to speak at a 'round table', a more open format where various researchers are invited to briefly speak about the status of a research programme, followed by ample time for questions. So I sent the subject of my presentation to Professor James Liebert, from the University of Arizona, the moderator of the round table. He replied a few days later asking if I could kindly explain my theme more precisely. How could

I object? Worried about saying too much, I ended up saying too little. Finally, I was given a mere five minutes to speak. I had hoped to have at least ten.

Despite its vagueness, our abstract started to circulate beyond the organising committee. The rumour about the discovery of an exoplanet grew. Some colleagues telephoned me to find out more. Among them was Steve Beckwith, from the University of Heidelberg, who had to speak about the cutting edge of exoplanet research at a conference in Catania a few days later. Naturally, he wanted to be aware of the latest developments in order to be as up-to-date as possible. I told him neither the name nor the properties of the planet, but I did confirm the news. In Catania, Steve rounded off his talk by announcing that a week later, at the Florence meeting, a Swiss team would announce the discovery of an exoplanet.

The secret was no longer a secret. Thanks to the curiosity of my colleagues, Professor Liebert agreed to give me more time than was originally planned. I would have forty-five minutes to speak. However, just because the end of my presentation was punctuated by vigorous applause, this didn't mean that I had convinced everyone. Questions flew furiously. This time, Didier and I finally had what we had waited for months for: a debate. Having heard about the news, Geoffrey Marcy, aka Dr Death, told me via email that he was off on the chase for 51 Peg's companion in order to either confirm or disprove its existence. Just for the record, my American colleague had dropped this star from his sample because a catalogue presented it as unreliable.

Ten days after his first message, I got the result of his analysis. Here is his message *in extenso*:

Dear Michel,
We have obtained 27 observations of 51 Peg, covering 4 days.
The velocities are sinusoidal, with an RMS residual of 2.5 m/s.
We find a period of 4.2 days, and amplitude of $K = 53 \pm 1$ m/s.
We understand that you found $K = 70$ m/s. I wonder if we differ
on K.

So your wonderful discovery is confirmed ! ! !
Congratulations again !
Alors, I will be asked by journals to comment on 51 Peg. Please
tell me if there is any problem with my stating to them that we
confirm the observational result.
I would be happy to send you our velocities. And I would surely
enjoy seeing your velocities.
Best regards,
Geoff

Encouraged by our result, Marcy and his colleague Paul Butler dived into the data they had accumulated over years of observations. Could it be that their spectrograph had detected short-period planets and that they hadn't noticed simply because they hadn't looked for them? A famous computer company provided them with an armada of powerful computers to accelerate the data reduction. The effort paid off. In January 1996, at a conference in San Diego, the American team announced the discovery of not just one, but two planetary companions, one in the Great Bear, 47 Ursa Majoris, the other in Virgo, 70 Virginis, which today is still one of the most massive yet found.

After lean times, the harvest was miraculous. Three planets in three months, what a magnificent result! Our science got back its self-confidence. This optimism has not faltered since. Today, planets are beating a path to the doorstep. They are ten a penny. Other worlds exist, there are thousands of them, an infinity of other worlds, just as the Greek Epicurus, and, a millenium later, the philosopher Giordano Bruno, had imagined.

2 Infinity and beyond

If the quest for exoplanets excites us so much, it's because it holds in promise the hope of maybe one day finding life elsewhere, life that was born in the light of another sun. It makes you dizzy just to think about it. What a shock it would be for humanity to discover that we are not alone in this Universe!

At the dawn of the third millenium, we're accustomed to talking about the vastness of the Universe. Infinity is almost ordinary. The latest generation of telescopes delivers images of the furthest jewels in the Universe to us. We're on first name terms with primordial galaxies, the first to have formed after the Big Bang. Bit by bit, we're putting together the history of the Cosmos. It's a tough job, but it can be done thanks to the progress in science since the beginning of the twentieth century and to the genius of people like Georges Lemaître, Alexander Friedman and Edwin Hubble, who showed that the Universe is not static, that it's expanding like a *soufflé*. The consequences of this discovery are nearly as infinite as the Universe itself. Because if it's getting bigger, then it must have been smaller when younger, it even had to have been born, from a 'singularity', as the experts say.

Today we estimate the age of our Universe to be about fourteen billion years. The initial singularity turned into a solidly built cosmos, which we would hesitate to call fully developed given that it seems guaranteed to expand further. In any case, on our modest scale of curious Terrestrials, it seems infinite. We know, thanks to telescopes, that behind the stars which form the constellations at night, there's more, much more, to discover. However, there was a time when this was not knowledge but pure speculation, or even heresy. You had to be very courageous to claim the unknown vastness of the Universe, or its infiniteness. Some even lost their lives for having added that in this vastness, there ought to be some earths similar to ours. They were philosophers, theologians and astronomers, and often supporters of the multiplicity of worlds, who refused to believe that our Earth was the only harbour of life in the Cosmos.

Since he influenced the whole of Western – including Arab – cosmology up to the end of the Middle Ages, Aristotle, who founded a school called the Lyceum, naturally steps first onto our historical stage.

ARISTOTLE AND THE FINITE UNIVERSE

When Aristotle (384–322 BC) died, he left behind a vast body of work. His scientific writings take prime position, though only a small part of Aristotle's work has reached us. Aristotle tried his hand at zoology, anatomy, physics and, of course, cosmology. He was not one of those astronomers who passed entire nights with their noses in the air, using their eyes to scan the starry ceiling in order to describe it down to the finest details. He preferred to imagine everything, to include the entire Universe, to portray it and to deliver the keys to how it works. Which is what he did in his *Treatise on the Sky* (*De coelo*). Aristotle's Universe is finite in space and infinite in time. It has two halves. On one side is the infralunar world, while on the other there's the supralunar world. As you can guess, the border between the two is our white Moon. Below it, the infralunar world includes the Earth and its atmosphere. Everything in this world is unpredictable, uncertain, approximate. Events never seem to repeat themselves in exactly the same way and they refuse to be trapped by simple and general laws of physics. It's disturbing to see how things and people change and die and also how life reconstructs itself in this world, even if it's never in quite the same way. With its disappearances and appearances, this infralunar Nature is like a magician who continually comes up with unexpected tricks.

Nothing is as ephemeral as this in the supralunar world. Everything is perfect, precise and stable. Nothing is created, nothing is

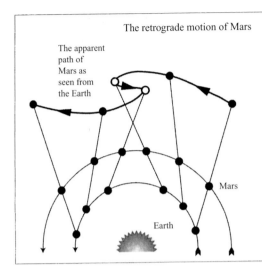

The retrograde motion of Mars

The apparent path of Mars as seen from the Earth

The retrograde motion of Mars – the apparent backwards motion (between the two white points) during its movement about the Sun – was a puzzle for ancient astronomers. Thanks to Copernicus and to the heliocentric model, it became clear that the apparently inconsistent movement of Mars is a sort of optical illusion created by the fact that the orbit of the Earth is smaller than that of the 'red planet'.

Mars

Earth

destroyed. In this world, the celestial bodies all follow circular paths in the sky. And the circle, as the Pythagoreans declared, is the epitome of perfection. Which is just as true for the sphere, the three-dimensional version of a circle. This is why Aristotle saw the Universe as a succession of spheres, nesting one inside another, like Russian dolls. The biggest of these supports the stars; it's the sphere of fixed objects. The smallest, which supports the Moon, marks the border between the infra and supralunar worlds.

In Aristotle's Universe the fifty or so spheres are not just a mental image. They really exist physically, and are made from a strange material, ether, which is said to be as transparent as crystal. Each of them has its own role. There's one that holds up Jupiter, another for Mars, another for Mercury, etc. In this strange but technically limited Ancient World, only five planets and the Moon were known. Add in the fixed stars, and you only really need seven spheres to describe the Universe. So why use more than fifty? For the simple and good reason that even if the sky is perfect, it's nevertheless prone to strange behaviour. It's enough just to think of planets' retrograde motion, a phenomenon in which planets give the impression of stopping in their celestial tracks, retracing their steps for some time, and then reverting to their normal routine movements. This deviates from the law of circles, the requirement of perfection. In addition, Aristotle imagined a full system of spheres, some of which carried nothing at all, but existed simply to turn in the opposite direction to that of the planets and thereby solve some of the apparent celestial contradictions.

Aristotle's cosmology was not really original. It borrowed a lot from other theories. For example, the idea of nested spheres is credited to Eudoxus (408–355 BC), a philosopher from Asia Minor (Turkey) who, like Aristotle, took the courses of the master of the Academy, Plato. It's due to this same Plato, by the way, that the Pythagorean thesis of the perfection of the circle was taken up and raised to the level of a universal principle. Plato taught whoever wanted to listen – and there were many – that it was absolutely necessary to 'make things look good', or in other words, to find the best way to explain these jumps in the sky while never contradicting the principle of the circle and the sphere.

EPICURUS AND THE INFINITE UNIVERSE

Aristotle's Universe was one answer among many to the Platonian decree. In spite of Aristotle's influence and the spread of his ideas, he didn't have a monopoly on theory. Others also juggled with the stars and planets, such as Epicurus (341–270 BC), whose philosophy was diametrically opposed to that of the master of the Lyceum. More known for his ethics than for his cosmology, Epicurus lived in a Greece that had lost a lot of its sparkle. The concern for the common good had given way to individualism. People looked less to the good of the community than to their own interests. The Epicureans advised living a life of pleasure, in moderation, but without depriving oneself. It was a wisdom that required you to free yourself from useless and irrational fears, like those of death, the gods or even Nature.

Epicurus' physics and cosmology supported this virtually therapeutic objective. They were inspired by atomism, a materialist philosophy, due, we think, to Democritus of Abdera around 400 BC.

According to atomism, the world is just an aggregate of tiny particles of matter, invisible to the naked eye, the *atomos*, literally 'indivisible'. Objects, people and even gods are all made of atoms. Only chance decides how these behave. Sometimes, they join together. Other times, they break apart. A stone can break up into pieces, split up into smaller bits, and finally disintegrate into powder. And just as there exist infinitely many atoms, the Universe is also infinite.

Epicurus faithfully followed in the path of the atomists, as can be seen in this extract from his *Letter to Herodotus:*

> The Universe is infinite. What is finite has an edge. An edge can only be perceived in relation to something exterior to that of which it is the edge: but the Universe cannot be perceived in relation to something which is exterior to itself, since it is the Universe; therefore, it does not have any edges and so it does not have any limits, and not having any limits, it must be infinite and not finite. We can add that the Universe is also infinite both in the number of bodies which it contains and in the size of the vacuum which is in it.

And a bit later:

> It's not only the number of atoms, it's the number of worlds which is infinite in the Universe. There are an infinite number of worlds similar to ours and an infinite number of different worlds ... Indeed, since the number of atoms is infinite, as we said earlier, they exist everywhere, their movement carries them even to the most faraway places. As well, again thanks to their infinite number, the quantity of atoms which can themselves serve as elements, or, in other words, as causes for a world, cannot be used up by the creation of a unique world, neither by that of a finite number of worlds, whether these are the worlds similar to ours or those which are different. Hence, there is nothing which prevents the existence of infinitely many worlds.

And finally, for a quite logical grand finale:

> One must admit that in all the worlds, without exception,
> there are animals, plants and all the other beings that we observe,
> because no one can show that one world is capable both of
> containing and of not containing the seeds of animals, plants and
> other beings that we observe; and that on the other hand, another
> world is absolutely incapable of containing such seeds.
>
> (Translation from Greek to French: Marcel Conche.)

Worlds, the Universe and even souls, everything is made of atoms and randomly mashed. It's no longer useful to fear either Nature or the gods. As for death, it's just a mental viewpoint. The *atoms* are dispersed, but they're not changed. Isn't this a sort of eternal promise? The Epicurean school was stunningly successful in its day, but it was defeated by time. Its materialist cosmology couldn't resist the power of the supporters of the sphere and the finiteness of the Universe.

PTOLEMY AND THE *ALMAGESTE*

We don't know much about Claudius Ptolemy (c. 100–180), except that he lived in Alexandria, Egypt, and worked at some time with an astronomer known as Theon of Smyrna (c. 125). It's difficult to believe that so little is known about someone who occupies such a major role in the history of astronomy. His major opus, the *Mathematical Syntax*, later renamed *Almageste* by the Arabs, represents, with its thirteen volumes, a fantastic compilation of Greek astronomical knowledge. As his work shows, Ptolemy was greatly inspired by his predecessors, like Hipparcos (190–120 BC), a very meticulous scholar to whom we owe the calculation of the solar year and of the Earth–Moon distance, as well as the first stellar catalogue detailing the positions of 850 stars.

Ptolemy's cosmology is impressively complex. Unswervingly geocentric, faithful to the Platonic principles, it juggles with an impressive number of spheres in order to best approach observed reality. And in fact, it succeeded better than any other previous theory. It describes the planets as forced to follow a small circular movement the centre of which itself follows a large circular movement. This formula, known as epicycles and already used by other astronomers, made it possible, while keeping to the perfect circle requirement, to more or less account for the apparent absurdities such as the retrograde hesitations of the planets and the strange shape of certain orbits. But the great innovation by Ptolemy was to remove the Earth from the centre of the Universe, or rather to shift it just a little in order to obtain a better description of phenomena such as the variations in the speeds of the planets. The *Almageste* is an ode to subtle geometric balances, a goldsmith's masterpiece for making things look good as Ptolemy wished, an 'ingenious sleight-of-hand', as Jean-Pierre Verdet has said.

The arrival of Christianity was soon to challenge Greek, pagan knowledge. The new religion was more interested in heavenly things than in material things. After all, if this Earth is sinful, what use is it to know if it's spherical or flat? Ancient knowledge, severely criticised by the Fathers of the Church, such as Saint Augustine (354–430), vanished. Symbolising this disappearance, Plato's Academy was officially closed in the year 529. A cloak fell on the sciences. Astronomy didn't entirely escape this dark destiny, even if it survived thanks to the work on calendars, a tool which even Christianity could not do without. This flame, feeble as it was, was, when the time was ripe, to light the way to the Renaissance.

While the West chose faith over knowledge, Islam practised both. In the eighth century, the Muslim empire had already expanded greatly, and the Caliph al-Mansur, based in Baghdad, ordered his scribes to translate the Greek works seized during his conquests. His son and successor, the famous Haroun al-Rachid, continued the task and it was during his reign that Ptolemy's *Almageste* was translated into Arabic. With the construction of a sophisticated observatory,

Baghdad became an astronomical city. Ptolemy was adored, his works were studied and debated by Arabic researchers. Measurements got better. Stellar catalogues became richer.

Western Europe, stuck in its religious dogmatism, nevertheless ended up by reviving Greek science, but not without provoking some serious bother in the Church. Ancient knowledge was disseminated from Spain, a country conquered by the Arab armies and where the Caliph was for a long time a promoter of tolerance. Educated Muslims, Jews and Christians met, discussed and exchanged ideas. Greek texts, with rich Arabic science added, were soon translated into Latin. At first the flow was just a drizzle, but then it became a downpour, thanks to the dedication of people like Adelard of Bath, Gerardo of Cremona and Dominicus Gudissalinus. During the twelfth century, these three alone were responsible for a hundred and sixteen translations, of which eighty-five concerned treatises of mathematics and natural sciences. These writings soon circulated in numerous church-funded schools and spread to the newly created universities. Even if the latter were mainly concerned with theological education, they still allowed some place for other knowledge.

Those who brushed against these ancient treatises were often obliged to carry out painful intellectual acrobatics. It proved very difficult to reconcile Greek reasoning with the mysteries of Christian faith. Nevertheless, certain scholars did attempt to bring the two together. Thomas Aquinas is undeniably the most famous of these. This Dominican was determined to use the resources of Aristotle's work to support his proof of the existence of God. He argued that the famous 'First Motor' of Aristotle, which generates the movement of the Universe without being moved itself by any other, was just another way of talking about the all-powerful divinity.

Not everyone thought like him, and the Dominican, who taught at the University of Paris, was soon subjected to the ire of a part of the Church outraged by such impertinence. But Thomas Aquinas resisted. He had been well trained. His teacher was none other than Albert the Great, who, in the thirteenth century, extensively campaigned for

Greek and Arabic philosophies to be disseminated in the Christian world. An open-minded, inquisitive man, convinced by the empirical approach, Albert the Great was very up-to-date with the sciences in general and with astronomy in particular. He even dared to tread in quite unorthodox waters:

> Do there exist many worlds or does only one exist? This question is without a doubt one of the most noble and exalting questions raised by the study of Nature.

Finally, the Church assimilated Greek knowledge, but not without having skimmed off the most diabolical provocations. It even did this in a way that consolidated its own foundations. In Aristotle's and Ptolemy's works, it found a scientific guarantee for its official cosmology: a perfect, spherical, centred, limited and hierarchical world which goes from Man to God. The celestial kingdom is perfect and immutable. It's naturally the reflection of God. But this beautiful arrangement didn't convince everyone. Dissident voices were soon heard, even in the ranks of the Church.

Nicolas of Cusa was born in 1401. He first studied law and mathematics, then he went to Cologne to study theology. Despite an exemplary ecclesiastical career, crowned by his nomination as Bishop of Brixen in 1450, the German was not that orthodox. He had very personal interpretations of certain Christian dogmas. He was possibly the first to break through the mediæval vision of the Cosmos by adding a few drops of infinity. In his book *On scholarly ignorance* he raised the question of why the divine power would have satisfied itself with making a closed Universe when it could do anything. There was something wrong, something that was illogical here. There had to be a huge Universe, so big, moreover, that it no longer has a physical centre, no longer a geographical middle. It only has a heart in a metaphysical sense: God.

After these first theological tremors, the Western world was about to go through a genuine scientific and conceptual earthquake, of which the epicentre was in Poland.

THE ARRIVAL OF HELIOCENTRISM

Nicolas Copernicus was born in 1473 in Torun, Poland. It was thanks to his uncle, a rich and powerful bishop, that as an adolescent he was able to enrol at the University of Kraków, which was famous for the excellence of its professorships in mathematics and astronomy. It's there that he first read Ptolemy's *Almageste*, which gave him a distinct feeling of unease. Why should the sky be so complicated? Wasn't there a simpler way of accounting for all its movements? Yes, there was, and Copernicus found it: you just have to put the Sun at the centre of the world and make all the other planets go around it, from the closest, Mercury, to the furthest, Saturn, going through Venus, the Earth with its Moon, Mars and Jupiter.

The heliocentric theory was born. Revealed in *De revolution- ibus orbium coelestium*, it was a lethal arrow shot into the heart of Aristotle's cosmology. If the Earth goes around the Sun and if the Moon goes around the Earth, how could you still accept Aristotle's claim that there are two worlds, two versions of physics, one above the Moon, one below? But Copernicus didn't totally break with tradition. He kept the existence of spheres and their solidity, as well as the need for celestial objects to follow circular motions, a theoretical conservatism that forced him to retain some of Ptolemy's archaic tools.

Thanks to his theory, Copernicus not only transformed the hierarchy of planets and the Sun, he also gave the Solar System dimensions hitherto never imagined, by stressing that, with respect to the vastness of the Cosmos, the circle traced by the Earth around the Sun is just a tiny point, nothing more. Geocentrism faded. Windows opened and the refreshing air of the Renaissance rushed in. Minds were set alight. Sometimes too literally because they ended up burned at the stake.

THE WORLDS OF GIORDANO BRUNO

Giordano Bruno was rebellious by nature. Consensus was not his thing. His independent character, bordering on insensitivity, was to cause terrible trouble for this Dominican. Uniquely, he was excommunicated three times and died in the flames of the Inquisition.

Bruno the opiniated entered this world at Nola, near Naples, in 1548. He was educated by various teachers and joined the Dominican Order of the Preachers at the age of seventeen. It didn't take the monks long to realise that they were dealing with someone with a mind of his own. They tried hard to keep him under control, but in vain. Young Giordano felt a wicked pleasure in upsetting dogmas, even the most symbolic ones. For example, he dared to doubt the Immaculate Conception and to disagree with the Holy Trinity. This was too much. Tired of his impertinence, the Dominican monks of Nola sent the young man home. If they hoped by this disciplinary measure to bring him back into line, they failed miserably. Bruno continued to defy orthodoxy. He read the works of Nicolas of Cusa, but also those of Lucretius (c. 98–55 BC), a Latin poet who supported the theses of the atomists and of Epicurean cosmology. Finally, he plunged into the works of Copernicus and became a fervent defender of heliocentrism.

After his ejection from Nola, Giordano Bruno went first to Naples and then to Rome, where the Dominican order continued to pursue him. He wandered for some time in northern Italy, continued into France in 1578 and then on to Calvinist Geneva. It might be thought that this land of reformation would be more receptive to his ideas but, in August 1579, Bruno was chased from Geneva by a Calvinist excommunication. He went back to France, first to Lyon, then Toulouse. His reputation grew, and for once it was positive. His use of mnemonic devices impressed crowds. Henry III was curious and invited him to court, where Bruno stayed for a few months before hitting the road again.

This time he chose England in the hope that there he would be able to defend his philosophy. But he got an ice-cold reception at Oxford, while in London, his neo-Copernican suggestions infuriated the conservative thinkers. Attacked from all quarters, Bruno defended himself by publishing three key works: *Ash Wednesday supper*; *On cause, principle, and unity*; *On the infinite universe and worlds*. This response didn't help at all. In the face of growing anger from his critics, he barely had the time to cross the Channel and head towards

Germany, where he fared hardly any better. In Marbourg, he was forbidden to teach philosophy. Continuing in his wanderings, he arrived in Prague in 1588. He hardly had the time to settle in before his provocative lectures earned him another excommunication, this time declared by the pastor of the Lutheran Church. Two years later, in Frankfurt am Main, he published three new works: *On the minimum*; *On the monad*; and *On immensity.*

Was he tired of fleeing? Was he homesick? Did he hope for a reconciliation with the Dominican order? We don't know, but Giordano Bruno returned to Italy at the invitation of a rich Venetian, Giovanni Mocenigo, who was interested in his mnemonic devices. However, for reasons which remain obscure, a few weeks later this benefactor decided to denounce his guest to the Inquisition. He was arrested on 22 May 1592. The list of charges was long. Bruno was accused of practising magic, of being interested in the occult and hermeticism, and, of course, of discussing Christian dogma. He was extradited to Rome and spent eight years rotting in Vatican gaols, eight years during which the Church asked him several times to renounce his heresies. In 1597, for instance, he was overtly asked to forget his claims about the plurality of worlds. Bruno unequivocally refused. He was convinced that the stars in the sky were suns around which planets full of life danced. Why would God have limited his power to create just one Earth? No, clearly, it's much more probable that he generated thousands and thousands and that each of these carries life. The inquisitors nearly choked themselves. How could one go to such lengths to deny the uniqueness of terrestrial creation? Pope Clement VIII eventually lost patience and ordered the execution of the heretic Bruno. Defying the judges, Bruno is supposed to have replied: 'You're no doubt more worried about giving this sentence than I am about receiving it.' He was burned at the stake on 17 February 1600 in the Campo dei Fiori.

While history made Bruno into a martyr for science, he was first of all a philosopher and a theologian. Like Nicolas of Cusa, he wondered about divine omnipotence: 'If God created everything that he was able to make, then the Universe could not be finite.' And if the Universe is infinite, that implies the existence of other Solar Systems:

> It is thus that the excellence of God is amplified and manifested by the greatness of his empire. It is not glorified just by one, but by countless Suns, not by just one Earth and a world, but by thousands of thousands, what am I saying? an infinity [of worlds].

And how can you doubt that some of these worlds are inhabited by beings 'similar or better than men'? Bruno saw life everywhere. It populates infinities. Even stars and planets have souls. This was a vitalism which contributed to costing him his life. The death of Bruno at the stake reminds us that dogmas go to extreme lengths to defend themselves. However, the revolution was under way, even if it sometimes looked like gentle reform, disguised as tradition and modernism.

TYCHO BRAHE AND THE THIRD WAY

Uraniborg. It sounds like a name from an action thriller. In Danish, it means 'celestial castle'. Its inhabitant was not an android, even if his nose had a metal plate in it, a legacy of a duel which nearly cost him his life. Tycho Brahe was the master of the house. He was born in 1546 in Knudstrup, three years after the death of Nicolas Copernicus. Coming from a well-off family, he had a full and integrated education starting from the age of thirteen at the University of Copenhagen. A year later, he was present at an event that would seal his future as an astronomer: a solar eclipse. He couldn't get over the fact that astronomers had successfully predicted it. Astronomy opened its arms to him and he fell in headlong. A few years later, as a reward for his success, King Frederic II of Denmark offered him the little island of Hveen, a few kilometres north of Copenhagen, where he could construct Uraniborg, his star city, his castle-observatory.

Tycho Brahe possessed an indisputable talent for designing observing instruments of excellent quality. Thanks to these, he

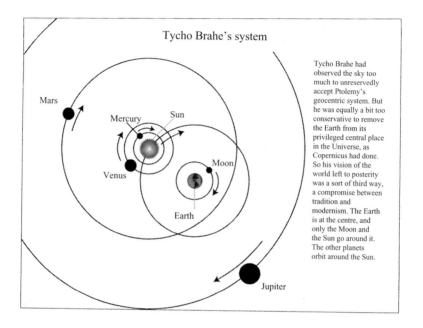

Tycho Brahe's system

Tycho Brahe had observed the sky too much to unreservedly accept Ptolemy's geocentric system. But he was equally a bit too conservative to remove the Earth from its privileged central place in the Universe, as Copernicus had done. So his vision of the world left to posterity was a sort of third way, a compromise between tradition and modernism. The Earth is at the centre, and only the Moon and the Sun go around it. The other planets orbit around the Sun.

improved the precision of measurements by a factor of ten. One evening in 1572, absorbed in his observations of the sky, he noticed a new star to the north-west of the constellation of Cassiopeia. It was incredibly bright, making Venus look pale by comparison. A comet, he thought first of all. Only comets are unpredictable enough to pop up where no one is expecting them. But this one was strange. Unlike other comets it didn't move. Night after night, Brahe found it at the same spot. Not knowing what it really was, he called it a *nova*.

Thus something new had appeared in the sky that Aristotle had said is immutable, where nothing can be created or disappear. The master of Uraniborg's surprises were not over. After several days, his *nova* started showing signs of growing fainter and a fortnight later it had definitively disappeared. Tycho Brahe didn't know it, but he had witnessed not the birth of a star but its death, its explosion.

The Danish astronomer had a particular interest in the sky that moves and changes. Before Brahe, the comets were considered to belong not to the sky but to the Earth, or more precisely to its atmosphere. Because with their foolish trajectories they upset the beautiful celestial regularity, Aristotle had associated them with a meteorological phenomenon, they were infralunar, otherwise they would have bumped into the crystalline ether of the spheres. Tycho Brahe didn't care about the Greek philosopher's arguments. His own observations contradicted Aristotle: comets travelled much further than the Moon. His conclusion was that the spheres were nothing but a mental image.

Was Tycho Brahe thus a revolutionary astronomer, a fighter against orthodoxy? Far from it, since his conservatism led him to reject Copernican heliocentrism. He was a fervent supporter of an Earth situated at the centre of the Universe. In contrast, his experience as an astronomer suggested to him that Ptolemy's system was not the right solution either. So he proposed a compromise, a hybrid theory where the Earth continued to occupy the centre of the Universe with the Moon and the Sun orbiting it, but where all the other planets turned around not the Earth, but the Sun. This is how Brahe's thoughts stretched in order to try to reconcile an ancient wisdom that was on its last legs with the wealth of modern discoveries. The Dane died in 1601, at the dawn of the seventeenth century which promised to be rich in scientific breakthroughs.

KEPLER'S LAWS

Kepler planned to become a pastor. Instead, he became an astronomer. Born in 1571 in Wurtemberg, Johannes Kepler had a difficult start in life. His family was poor, his father was absent and he was often ill. Luckily for him, the Duke of Wurtemberg, an enlightened man, gave scholarships to promising children in difficult circumstances. The young Kepler was granted one of these. At the end of his studies, he was appointed as a lecturer in mathematics in a college in Gratz (today Graz, in Austria). Copernicus' work fascinated him. In March 1596, he published his first astronomical document, the *Mysterium cosmographicum*, in which he openly stated his opposition to the dogma of celestial spheres. This publication had quite some success

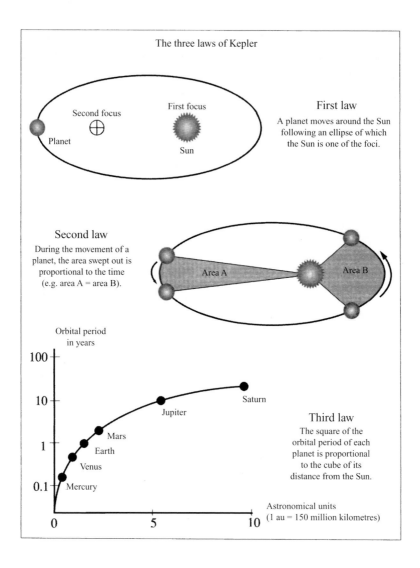

The three laws of Kepler

First law
A planet moves around the Sun following an ellipse of which the Sun is one of the foci.

Second focus
Planet
First focus
Sun

Second law
During the movement of a planet, the area swept out is proportional to the time (e.g. area A = area B).

Area A
Area B

Orbital period in years

100
10
1
0.1

Mercury
Venus
Earth
Mars
Jupiter
Saturn

0   5   10

Third law
The square of the orbital period of each planet is proportional to the cube of its distance from the Sun.

Astronomical units
(1 au = 150 million kilometres)

and gave its author enough of a reputation for him to join Tycho Brahe's group. The latter had left Uraniborg after a quarrel with the Danish sovereign. He had set himself up in Prague, where he welcomed his new assistant. He asked him to work on the chaotic movements of the planet Mars, certainly one of the most arduous problems of celestial mechanics. Even after the death of Tycho Brahe, Kepler continued his Martian investigations. He used all the mathematical tools at his disposal to tame the rebel planet, a task requiring several years that led him, to his utmost reluctance, to give up the requirement that celestial movements must be described by a circle. He gave the planetary orbits the shape they deserve, the ellipse, and in the same breath stated the three laws which made him famous:

(1)   A planet moves around the Sun following an ellipse of which the Sun is one of the foci.

(2)   During the movement of a planet, the area swept out is proportional to the time.

(3)   The cubes of the lengths of the major axes of the planets are proportional to the squares of their periods.

Johannes Kepler didn't escape the debate about the size of the Universe. If he hadn't read the works of Giordano Bruno, he had at least heard about them, since he referred to them in his book *Opera omnia*. But their concepts conflicted. Kepler doubted that the Universe is infinite as proposed by Bruno. Because if that were the case, then stars would have to be uniformly spread throughout space. Whereas a glance was enough to show the irregularities in their distribution. An observer placed on a star other than the Sun would not have the same image of the Universe as us. Sure, it was possible that some stars, due to their small size or their large distance, escaped from our view and that these filled the irregular spaces. But this was nothing but pure speculation. Infinity was the business of metaphysics and not of science.

GALILEO GOES FURTHER

At the same time, in a pretty corner of Italy, a scholar wrote the following lines:

> These are great things, in fact, that I propose in this brief treatise for the contemplation of those who study Nature. Great by the excellence of the subject itself, by their novelty since no one

knew about them in any previous epoch, and also by the instrument thanks to which they have manifested themselves to our senses.

It's a great thing, surely, to have added to the multitude of fixed stars already discovered up to now by the simple view of the naked eye, different and innumerable stars, never perceived before our epoch, and to expose them to our viewing in numbers more than ten times greater than that of the ancient stars already known.

Thus began *Sidereus nuncius* (*The messenger of the stars*), Galileo's major publication. It was 1610, and the Florence-born Galileo was 46. At the time, he was professor at the University of Padova, which belonged to the Republic of Venice. His refracting telescope impressed the Venetians. Thanks to a few glass lenses inserted in a tube, Galileo showed the sky in a totally new way. Stars appeared as if by magic, while the Moon revealed the incredible diversity of its mountains, its seas and its craters. It was far from the Moon as smooth as a mirror described by Aristotle. By showing it in all its ruggedness, Galileo elevated it to the rank of a new 'world', to the rank of a planet. And he didn't stop there.

His telescope was really doing wonders. It soon let him discover four small brilliant points turning around Jupiter. But rather than call these satellites, the Italian scholar preferred talking about 'Medicean planets'. In total, thanks to a few lenses inserted in a small tube, Galileo increased the Solar System by five new worlds and a new system, that of Jupiter.

All these discoveries came to reinforce the heliocentric theory of which the Florentine declared himself an ardent supporter. His points of view, which were a little too vehement for the liking of the Church, earned him, moreover, some serious problems with the Inquisition. He was condemned on 22 June 1633 and had to publicly renounce his heretical convictions. But it was too late: the core of his work had already been disseminated.

Copernicus, Kepler, Galileo and Tycho Brahe were four powerful minds who paved the way to a fantastic scientific revolution. They overthrew the traditional order of the Cosmos. However, each in his own way remained anchored to that tradition. It's not that easy to totally cut off the anchor. So there was still a lot to accomplish by geniuses in need of breakthroughs. One of them was to stupefy his contemporaries by delivering the image of a Universe which was subject right down to its obscurest corners to the same physical law, the force of gravitation.

ISAAC NEWTON'S GRAVITY

Isaac Newton (1642–1727) was a strange man. Born in Woolsthorpe and raised by his grandmother after the death of his father, he was a stormy character to say the least. His taste for solitude rarely allowed exceptions. One evening when he had people over for dinner, he left his guests for a moment to look for some wine. After several minutes, the guests, worried about his prolonged absence, went to look for him and soon found him, standing at his desk filling pages with calculations. His extraordinary intellectual power permitted him to combine the breakthroughs of Galileo and Kepler. He understood that the movement of the planets followed the same laws as the fall of bodies. Heaven and Earth united under identical mechanical codes, governed by the same physical laws. The two worlds of Aristotle went into the dustbin. Newton took over. Thanks to him and his equations of universal gravitation, men would one day walk on the Moon.

But imagination often proceeds faster than technical progress: imagination such as that of Bernard Le Bouyer de Fontenelle (1657–1757), a Frenchman of letters passionate about astronomy and unabashedly audacious. He was one of those who thought that life is not a terrestrial privilege, that there exist other worlds around stars other than the Sun. His thesis was daring for his time, even if the Inquisition was over.

THE BEGINNINGS OF SCIENCE FICTION

Fontenelle died a centenarian and famous. His book titled *Interviews about the plurality of worlds* (1686), in which a man interviews a beautiful duchess on the secrets of the sky, was a bestseller and went through several editions. From this point on, astronomy was no longer the private affair of a circle of insiders. It had become a topic for general discussion. Fontenelle was not the first to talk about extraterrestrial life. Kepler had done so before him in his strange book, *Somnium* (*The dream*), where poetic lyrics mix with the astronomical knowledge of the time. More down-to-earth, the Dutchman Christiaan Huygens (1629–1695), discoverer of the secret of Saturn's rings and master of physical optics, put forward in his *Cosmotheoros* (*The spectator of the universe or Hypotheses on celestial worlds and their movements*) the thesis of the plurality of worlds:

> Someone who accepts Copernicus' opinion, who makes our Earth a planet like any other, pulled around the Sun and illuminated by it, such a person can reasonably believe, even if this seems rather daring, that the other planets [...] have inhabitants just like the Earth.

He also exploited this to settle an argument about anthropocentrism which continued to cloud people's minds:

> What obliges me to believe that in the planets there is an animal which reasons, is that without that the Earth would have too great advantages, and would be too raised in dignity above the rest of the planets, if only she had an animal so strongly raised above the others.

And if humanity alone was not the only being gifted with reason in the Solar System, the chances are that the Universe has yet more marvels:

> Let's not hesitate, us, to admit with the principal Philosophers of our time, that the nature of the stars and that of the Sun is the same. Which implies a conception of the world much more grandiose than that which corresponds to previous more or less traditional views. Since what stops us now from thinking that each of these stars or Suns has Planets around it and that these planets in turn are endowed with Moons?

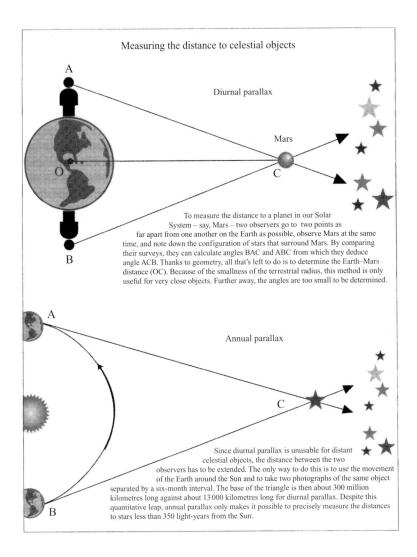

Measuring the distance to celestial objects

A

Diurnal parallax

Mars

O

C

B

To measure the distance to a planet in our Solar System – say, Mars – two observers go to two points as far apart from one another on the Earth as possible, observe Mars at the same time, and note down the configuration of stars that surround Mars. By comparing their surveys, they can calculate angles BAC and ABC from which they deduce angle ACB. Thanks to geometry, all that's left to do is to determine the Earth–Mars distance (OC). Because of the smallness of the terrestrial radius, this method is only useful for very close objects. Further away, the angles are too small to be determined.

A

Annual parallax

C

Since diurnal parallax is unusable for distant celestial objects, the distance between the two observers has to be extended. The only way to do this is to use the movement of the Earth around the Sun and to take two photographs of the same object separated by a six-month interval. The base of the triangle is then about 300 million kilometres long against about 13 000 kilometres long for diurnal parallax. Despite this quantitative leap, annual parallax only makes it possible to precisely measure the distances to stars less than 350 light-years from the Sun.

B

MEASURING THE SKY

Freed from the grip of the Aristotelian spheres, the Cosmos expanded and expanded again. On top of this, techniques got better and observing instruments became more powerful. People were continually discovering more stars in a sky which seemed never lacking in resources.

At the end of the seventeenth century, two French astronomers, Jean Richer and Jean-Dominique Cassini, worked to determine the distance between the Earth and the Sun – they evaluated it as 146 million kilometres. They were wrong by only 3.6 million kilometres, a feat for their time. In fact, the technique that they used, that of parallaxes, had existed for ages – the Greek Hipparcos described the principle – but the necessary mathematical tools hadn't been available.

To understand what a parallax is, it's enough to hold your arm in front of you, with your thumb lifted, and to look at your thumb with each eye alternately. You'll see it shift with respect to the background scene (the wall of the room, for example). This movement comes from the separation of our eyes which makes each of them have its own viewpoint on the world. This separation, the size of which is measurable, is like the base of a triangle the other two sides of which join up at the distance of your thumb. As the angle of the vertex (at your thumb) can be deduced from the shift of the object relative to the background, you can therefore calculate the length of the two sides of the triangle using trigonometry and then obtain the distance between the observer and the object being viewed, that is, between your eyes and your thumb.

If you want to calculate the distance between the Earth and another planet or between the Earth and the Sun, using your thumb is not good enough. Instead, your eyes are replaced by two astronomers who observe the sky at two different places on the globe. The greater the distance that separates them, the easier is the operation. Both look at the celestial object of interest and note down its position with respect to stars in the background. By comparing the two views, you can deduce the angle, which in turn makes it possible to determine the length of the long sides of the triangle and finally the distance to the object of interest. Because the Sun shines and complicates measurements Richer and Cassini used Mars as a sort of relay for calculating the parallax of the Sun. Edmond Halley (1656–1742), the comet man, carried out a similar calculation using Venus as an intermediary. He thus obtained an Earth–Sun distance of 151.5 million kilometres.

The accumulation of measurements started to lead to an understanding of the vastness of the Solar System. And the stars? How far are they? The first to obtain a good estimate was the Swiss Jean-Philippe Loys de Chéseaux (1718–1751). He was of a frail constitution as a child. Maybe that explains why very early on he began studying the sciences. His gifts were astonishing and his curiosity insatiable. Mathematics, astronomy, theology, ancient languages, he juggled with knowledge. As a young adult, he launched into particularly arduous problems for the epoch. Not happy with inspiring himself from the works of Newton and Kepler for predicting the future trajectory of the comet that he observed in 1744, he tried to crack the secret of the blackness of the night. Why is the night black when logically all the light from the stars ought to make it shine?

Loys de Chéseaux wanted to resolve this question, which others had discussed before him, by quantitative means which themselves required that certain parameters such as the distances to the fixed stars had to be known. The young Swiss observed that the planets are about as bright as the brightest stars. He then assumed that each of the stars was a sun. Since the distance from the Sun to the planets was known approximately, it was possible to calculate the reduction in light intensity with distance. Of course, a planet only reflects light, it doesn't create it like stars, a detail that was not overlooked by Loys de Chéseaux. By churning through the calculations, he came to a result which was in light-years correct to an order of magnitude. Who knows what the young prodigy could have given to science if he had not succumbed to illness when he was only thirty-three?

We have to jump to the first half of the nineteenth century to obtain more accurate measurements. In 1838, the German astronomer Friedrich Bessel, professor of astronomy at Königsberg, declared that the star 61 Cygnus, one of the brightest in the sky, was located about 11 light-years away, i.e. 104 060 billion kilometres. Once again, it was the parallax technique that made it possible to make this discovery. This time there was an important difference, however: the German master of astrometry had used annual parallax. Rather than choose two distant points on the globe – no distance on Earth is big enough to make it possible to obtain an angle which is easily measurable – Bessel changed the scale and chose to use the diameter of the terrestrial orbit. He looked at the star in the spring and again six months later, in autumn. Some 300 million kilometres separate the two viewpoints. This is enough to measure the distance to the stars, or at least, to the closest ones.

While 11 light-years is a big distance, the odds were that there were numerous objects further away still. Maybe this would also be the case for the strange white smudge which streaks across the night sky. Greek mythology saw in it drops of milk spread from the favours of Hera, the wife of Zeus. Hence, the name of the Milky Way. Differently, some ancient scholars, like Democritus, the atomist from Abdera, claimed during their epoch that this cosmic cloud was in fact a concentration of stars that were too far and so too small to be distinguished individually. While correct, this intuition was unverifiable at the time. Democritus would have had to have had an instrument with which he could see the stars better. This privilege was granted to Galileo some two thousand years later, and made him write:

> What has been given to us to observe is the nature or better the matter of which the Milky Way is constituted [...]. The galaxy is nothing other, in effect, than a cluster of countless stars disseminated in little bunches: in whatever region one points the refractor, it immediately shows a view of a considerable number of stars, of which many can be seen big and distinct; but the multitude of small stars remains completely undiscernable.

In the eighteenth century, the Englishman Thomas Wright, surpassing Galileo in audacity, asserted in his book *New hypothesis on the Universe*, published in 1750, that the Milky Way is like a great disc of stars, among which the Sun is found. Seized by a bout of lyricism, Wright then wrote:

> What an extraordinary spectacle is deployed in front of our eyes, what unconceivably vast and magnificent power is revealed by such a state of the world? Suns bunched with other suns,

according to our feeble senses, unmeasurably distant from one another; myriads and myriads of habitats similar to ours, populating the infinite, all subject to the same will of the Creator; a Universe of worlds all covered by mountains, lakes, seas, grasses, animals, rivers, rocks, caves, and to share infinity with infinite beings with which his omnipotence could give an eternal life full of variety.

THE WALTZ OF THE GALAXIES

This text grabbed the attention of a young intellectual called Emmanuel Kant (1724–1804). At the beginning of his fantastic career, the great thinker from Königsberg was not yet totally absorbed by philosophy. Excited by science, he came one day in 1751 on an article summarising the theses of Thomas Wright. The tale impassioned him, just like the writings of the French mathematician Pierre Louis Moreau de Maupertuis (1698–1759) on the nebulae. Ideas were shuffled. Kant soon brought them together in a book titled *A general history of nature and theory of the sky*. Its principal protagonist was universal gravitation. It was this that modelled all celestial objects. In the beginning, the world was just stuff scattered all over the place, an indescribable chaos. Gravitation put things into place. Clouds of matter formed, which gave birth to the stars. Then, like a sheepdog, gravitation collected stars together into flocks, galaxies. In France, because the genius mathematician Pierre Simon de Laplace (1749–1827) had announced a similar idea to that of Kant, the hypothesis thereafter carried the name of the two scholars. Their sky is fabulously deep. In their hands, the Universe became an ocean without end, dotted with countless island galaxies, the nebulae.

What a perspective! How dizzying! But does any of that have any scientific relevance? The astronomer William Herschel (1738–1822), whom we'll meet again shortly, asked himself this question. As good an astronomer as a mathematician, he designed the largest and most powerful telescopes of his time. He saw further than anyone before him. He first followed Kant's theses but then moved away and reduced the entire Universe, or at least that which is observable, to just the Milky Way. For Herschel, the nebulae are not outside, but inside the Milky Way. This vision of the Cosmos dominated the nineteenth century. One had to wait for 1920–1930 and the work of the famous American astronomer Edwin Hubble in order to understand that the Andromeda nebula, M31, which is visible to the naked eye, is a fantastic concentration of stars located outside of our Galaxy.

Today, we know that Andromeda is floating about 2 million light-years from the Sun. This makes it a very close neighbour on the scale of a Universe which must be about 12–15 billion light-years in radius. Billions of galaxies inhabit this immense cosmos. You can hardly imagine the extraordinary number of stars that this represents. Just as it's difficult, even if it's less important, to represent the number of planets that orbit around these distant Suns. There was a time when no one would have bet on the existence of a planet further than Saturn. But in less than two hundred years, three new celestial objects were added to the Solar System. Planet chasing is an ancient art.

3 New arrivals in the Solar System

From the most ancient times right up to the sixteenth century, astronomers knew of only five planets (Mercury, Venus, Mars, Jupiter,
Saturn) as well as the Earth, which was delicately nested at the heart
of the Universe. For a very long time, this geocentrism constituted
the dominant vision of the Western world, right up to the day when
Nicolas Copernicus put the church back in the middle of the village
and the Sun in the centre of the Solar System.

The Earth no longer reigned at the centre of everything. It became an appendage of the day star, a planet like any other. The Universe was turned topsy-turvy. And in addition to this, there were
Galileo's Medicean planets. The Solar System was taking shape, and
there was nothing to stop the telescope from finding new worlds. Anything was possible, except, perhaps, finding a planet beyond Saturn.
It was still thought that the 'lord of the rings' ended the world of the
planets and that after it there was nothing but stars. It took nearly
two centuries for this model to be laid to rest.

HINTS OF URANUS

The German William Herschel was born in 1738 in Hanover, Prussia.
His family was a big one. His parents, Isaac and Anne, had ten children.
Four died young. The other six were raised to the regular rhythm of
scales played by their father, an oboe player in the military band of the
Guards of Hanover. Isaac was a cultivated and inquisitive man. In his
house, mealtimes were often animated, pretexts for great discussions,
which was not quite to the liking of Anne, who was much more severe
than her husband.

In the spring of 1753, at the age of only fourteen, William, having learned the oboe and the violin, joined his father's band. Three

years later, the band went off to England, where it stayed for nearly ten months. This was a result of the Seven Years War, in which Prussia and England fought France and Austria. The young Herschel brought back from his visit to England a thorough knowledge of the language of Shakespeare. But in Germany, the pace of events quickened, with France gaining ground. Even as a musician, William suffered the torments of military campaign life. He couldn't handle it. Since he knew people in the right places, Isaac was able to have William released from the band, which saved him from falling into French hands a few months later. William and his older brother Jacob preferred to go into exile in England, where they lived by their music. When the French were finally defeated two years later, Jacob returned home, but William chose to stay put and make his life in England.

He felt comfortable in his role as a music teacher and made a good living. Enough in any case to invite his sister Caroline to join him. He thus freed her from the thankless task of serving a difficult mother, who agreed to let her daughter leave only on the condition that William paid for a housemaid in compensation. He did this and left for England with his sister, who started a new, freer, happier life, though not necessarily with any fewer constraints.

At first, Caroline worked on her voice which was said to be quite beautiful. Soon, she was even giving performances. For his part, William became more and more interested in astronomy. He arrived there by a strange detour. It was when studying musical harmony that he had first discovered its relation to mathematics and he developed a passion for handling figures and equations. Via mathematics, the world of physics was opened up to him and he became crazy over one of its fields, optics, which then finally led him quite naturally to astronomy.

Herschel started with theory. This was less by choice than by default. Doing astronomy required having a telescope, and these instruments were still very expensive at the time. In the end, the young musician decided to make one himself. He happened to meet an amateur optician who was tired of his hobby and who agree to sell his materials to Herschel for a modest sum. William was now equipped for making a telescope. It didn't take him long to realise that melting and polishing glass lenses was not an activity that could be carried out easily. So he changed his line of attack and decided to construct a refracting Newton telescope, an instrument that depends not on lenses, but on mirrors. Even though the latter are easier to make, Herschel spoiled several dozen before obtaining a satisfactory result. It was an exercise in patience and stubbornness, but for this enthusiast, used to playing scales, nothing was impossible.

His first telescope had a 15-centimetre primary mirror. This was pretty modest compared with the 1.40-metre giant that he achieved a few years later and with the instruments that were to make his reputation, those which he sold to kings and princes in the remotest corners of the world. Bit by bit, Herschel added to his celestial objects. Each time that the weather was good enough, he passed his nights with his eye riveted to his telescope. The rest of the time, when clouds, rain or fog got in the way, he went to his workshop to melt and polish his mirrors.

Rather than observing planets, which was, however, very much the in thing at the time, Herschel preferred to look at stars and nebulae. Soon, he identified the latter as stellar concentrations. Through his frequent observations Herschel gained in experience to the point of matching professional researchers. Between music and astronomy, he didn't have a moment of respite. Luckily, he could rely on a devoted assistant, his sister, who would even spoon feed him when he was too busy polishing a mirror to take time to eat.

Very soon, Caroline was also helping William in his scientific endeavours; he gave her daily lessons in algebra and trigonometry. The young woman would no doubt have preferred to continue singing, but she acquired a taste for astronomy and made a brilliant career out of it. With eight new comets to her credit, the addition of five hundred and sixty stars to John Flamsteed's renowned stellar catalogue and the composition of an original catalogue of two thousand five hundred nebulae, Caroline was decorated by the Royal Society and received from it an annual stipend of fifty pounds.

One evening in 1779, when he had put his telescope out on the street, William Herschel was asked by a passerby to be allowed to observe the Moon. Friendly by nature, Herschel willingly agreed. He didn't know at the time but the passerby was none other than William Watson, a member of the Royal Society. Impressed by the meticulousness and the knowledge of the amateur astronomer, the eminent scientist encouraged him to submit his work to members of the Society. A year later, Herschel presented his first paper to the Royal Society: it concerned the height of the mountains of the Moon. But these lunar investigations were not able to camouflage his increasingly more marked interest in stars. It was while observing stars that, on the night of 13 March 1781, he noticed a funny object at the boundary between the constellations of Taurus and Gemini.

The object was too fuzzy to be a star. Herschel hesitated. Was it a comet or a nebula? He continued his observations the following nights and noticed that the object had moved. As nebulae are always stationary against the sky background, it could only be a comet. At no moment, it seems, did the astronomer imagine that the object could be a planet. Yet, several clues should have put him on the right track. First of all, the object crossed the constellations of the Zodiac like all other known planets. Also, if it was really a comet, it should have shown the usual long, feathery tail. But it didn't have one. And to top it off, while other comets are always characterised by very elliptical, elongated and off-centre orbits, this one followed a circular orbit, like the planets. It was some time before Herschel understood the real significance of his discovery.

The English Astronomer Royal Nevil Maskelyne was the first to openly talk of a seventh planet. Soon other observations supported his claims. Laplace, who worked on the orbit of this object, also declared that it closely resembled that of a planet. Herschel ended up by agreeing with this opinion and exercised his right to name his discovery.

He proposed *Georgium sidus*, literally 'George's star', in honour of the king of England. It was very diplomatic, but this initiative didn't appeal much to the astronomical community, which held firmly to the tradition of names from Greek mythology. The German Johann Bode suggested Uranus, who, in the Greek pantheon, is none other than the father of Saturn. The name stuck, except with Herschel, who continued to talk of *Georgium sidus* right up to his deathbed.

Thanks to this discovery, the music teacher achieved celebrity status and honours. At the age of forty-three, he was decorated by the Royal Society at the same time that he became a member of that venerable assembly. The king awarded him an income of two hundred pounds per annum. From then on he was able to dedicate himself full time to astronomy. The budding musician metamorphosed into an adult astronomer. In 1788, at the age of fifty, William Herschel married Mary Pitt. From this match was born a son who was educated mostly by Caroline Herschel. The son, named John, in his turn became an astronomer and played a major role in the field of planetary discoveries.

After the discovery of Uranus, anything seemed possible. If there existed a seventh planet, why shouldn't there be an eighth, a ninth and even a tenth? Astronomy was in a tizz. If only one knew where to look, what clue to follow? The sky is so vast! Yes, the sky is vast, but Newton had left his colleagues something to make the quest easier. Celestial mechanics, augmented by the law of gravitation, constitutes an instrument of formidable efficiency. It's like a mathematical telescope.

NEPTUNE IS REVEALED

Neptune, the eighth planet of the Solar System, could have been discovered by traditional optical methods. Indeed, Galileo saw it on 8 December 1612, but listed it just as an eighth magnitude star. (Magnitudes describe the brightness of a celestial body: the higher the magnitude, the less bright. Sirius, the brightest star in the sky, has a magnitude of −1.4. The naked eye cannot see stars that are beyond a

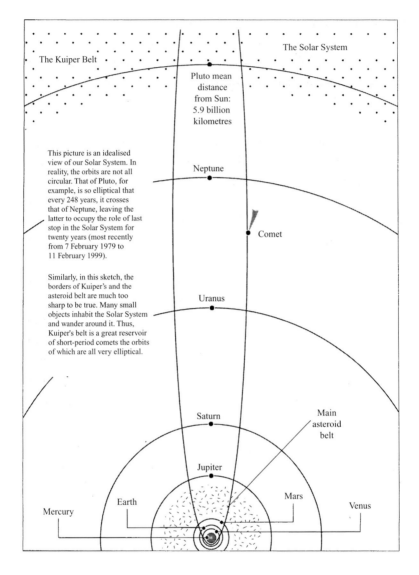

The Kuiper Belt

The Solar System

Pluto mean distance from Sun: 5.9 billion kilometres

This picture is an idealised view of our Solar System. In reality, the orbits are not all circular. That of Pluto, for example, is so elliptical that every 248 years, it crosses that of Neptune, leaving the latter to occupy the role of last stop in the Solar System for twenty years (most recently from 7 February 1979 to 11 February 1999).

Similarly, in this sketch, the borders of Kuiper's and the asteroid belt are much too sharp to be true. Many small objects inhabit the Solar System and wander around it. Thus, Kuiper's belt is a great reservoir of short-period comets the orbits of which are all very elliptical.

Neptune

Comet

Uranus

Saturn

Main asteroid belt

Jupiter

Mercury

Earth

Mars

Venus

magnitude of 6.) A month later, he saw it again beside another star. The following day, he continued his observations and noted that the two objects were a bit further apart than they had been the day before. So one of them was moving. And if it moved with such a speed, it couldn't be a star. So? So, nothing. The Italian astronomer wrote nothing else. Why is a mystery. And he wasn't the only one to pass over this treasure. On 8 and 10 May 1795, a Frenchman, Joseph Jérôme de Lalande, in his turn saw Neptune without noticing anything special. On 14 July 1830, John Herschel, William's son, also noticed it without identifying its planetary nature. Two planet discoverers in the same family would have been remarkable. But Neptune was finally revealed to the world thanks to mathematics and mechanics.

At the end of the eighteenth century, several researchers began the task of making precise calculations of the different planetary orbits. As each of the paths is necessarily influenced by the presence of the other members of the Solar System, you need to take into account the masses of the latter in order to obtain a reliable result. The Earth would certainly not follow the same route if it went around the Sun alone, without the influence of the other planets.

Yet, in the case of Uranus' orbit, something was wrong. After having painstakingly taken into account all the gravitational influences that it was subject to, the experts were left with an inexplicable remainder. There was a shift of 4 arc minutes, i.e. an eighth of the apparent diameter of the Moon, between the calculated position of Uranus and its observed position. How could this be explained? John Couch Adams, a student at Cambridge University, had the germ of an idea and wrote these few words in July 1841:

> I've decided at the beginning of this week to enquire as soon as possible, after having passed my degree, on the irregularities of Uranus' movements of which the cause has not yet been found; and this would be to find out if the movements could be due to the presence of a yet unknown planet; and if possible to determine its properties, the elements of its orbit in order to discover it.

As well as the disovery of Uranus, another clue encouraged astronomers to look for a possible eighth planet. This was the 'Titius–Bode' law, which shows that the positions of the planets in the Solar System follows an algebraic sequence.

In 1766, a mathematics professor at Wittenberg, Johann Daniel Tietz, known as Titius, got the idea of translating an important work of the time, *Contemplation of nature*, in which the Genevan naturalist Charles Bonnet declared his admiration for the apparent regularity in the distribution of Solar System objects. Intrigued by this remark, Titius decided to verify mathematically whether this pattern followed some law. It seems it does. You just have to divide the distance from the Sun to Saturn into a hundred units and to look at the location of each planet on this scale. Mercury, the closest to the Sun, is at 4 units, which can also be expressed as $0 + 4$. Venus, the second, is found at 7 units, i.e. $3 + 4$. Now the Earth: it's at 10, i.e. $6 + 4$. Mars is at 16 units ($12 + 4$), and Jupiter is at 52 ($48 + 4$). Finally, Saturn is at 100 units ($96 + 4$). So there is clearly a geometrical sequence in which each term is the double of the preceding one: 3, 6, 12, (24), 48, 96. To add to the impact of this law, Titius discovered that the satellites of Jupiter follow a similar sequence. This was undoubtedly the work of God. It remained to know why the divine architect had not created a planet at the level $24 + 4$, between Mars and Jupiter. Was it forgetfulness? Can God forget? No, since God is omnipotent and omniscient. So it was clear, there couldn't be a hole at $24 + 4$. There had to be objects there that astronomers hadn't yet detected, maybe moons lost from Mars and Jupiter. Titius was so convinced of his theory that he added it to his second translation of the work of Bonnet.

Years later, another German, Johann Bode, decided to follow up the work of Titius. Rather than the hypothesis of lost moons, he preferred that of a real planet, and concluded that all that remained was to find it. His optimism sometimes puzzled his peers, who asked themselves how such a close celestial body could have escaped detection for so long. But the doubts faded with the discovery of Uranus, whose position agrees with the algebraic sequence. So, in September 1800, six astronomers met in Lilienthal, not far from Hanover, to decide on a strategy for looking for this hypothetical planet located between Mars and Jupiter. Reinforcements were needed and several hunters joined the chase. As all the known planets pass through the constellations of the Zodiac, in the plane of the Solar System, it's there that efforts were concentrated. The horizon was divided into twenty-four zones and each zone was allotted to an observer. A list of the best astronomers was drawn up, and letters were sent to the four corners of Europe.

One of these was addressed to the Sicilian priest and mathematician Giuseppe Piazzi, of the Observatory of Palermo. By a twist of fate, he made a crucial discovery before even receiving the appeal. Several years earlier, the Italian had committed himself to a bold exercise in stellar accounting. So his eye was already trained. The object that he noticed on the night of 1 January 1801 in the constellation of Taurus was absent from all of his catalogues. It was very faint, ten times less luminous than the planet Uranus. But it wasn't a star, because observations over several nights showed that it shifted against the background sky. Piazzi officially identified it as a comet, even though he confided to some friends that it could well consist of a more massive object, maybe even a small planet.

A few weeks later, Bode learned about Piazzi's discovery. For him, there was no doubt that the Italian had found 'his' planet. This needed confirmation. Alas, it was at that moment that Piazzi's object chose to slide behind the Sun. Other astronomers tried to find it but couldn't. What made matters worse was that Piazzi hadn't been able to observe the object long enough to calculate its orbit. The data were just too few. The astronomer's impatience grew. Then a twenty-three-year-old German scientist focused on the problem. Carl Friedrich Gauss was fascinated by this invisible object. He decided to calculate its orbit. Not bothered by the lack of data, he invented a new calculation method which got around the problem. He sent his results to the Hungarian Baron Franz Xaver von Zach, one of the participants of the Lilienthal meeting. Realising that he had in his hands extremely reliable coordinates, von Zach rushed to the observatory and located the object within a half a degree of the position predicated by Gauss, and at a distance of 2.77 astronomical units, whereas the Titius–Bode law predicated 2.8 au.

The only sour note in this concert of successes concerned the size of the planet, from then on named Ceres. Its diameter was just over 900 kilometres, which was why, despite its proximity, it was so difficult to find. But was it really a planet? Debate was raging when an announcement came. On 28 March 1802, the medical doctor and amateur astronomer Heinrich Wilhelm Olbers, a member of the Lilienthal group, had discovered a new celestial body, very similar to Ceres. He called it Pallas. The astronomers were bewildered. They had thought, like Bode, that they would find a decent-sized planet, and instead they had found two small objects. The family of small objects grew bigger yet with the discoveries of Junon in 1904, by Carl Ludwig Harding, and Vesta in 1807, by Olbers. For Olbers, there was only one way to explain this profusion of small objects: they had to be the remains of a planet that for various reasons had been blown to bits in the distant past. William Herschel proposed calling these remnants asteroids. The name stuck. We've now counted thousands of them between Mars and Jupiter which together form what we call the asteroid belt. Despite their number, they only represent a mass equivalent to a thousandth of that of the Earth. Rather than planetary debris, they seem to be the remains of the disc of matter which in the infancy of the Solar System gave birth to the planets.

The discovery of the asteroids proved that chance observation was no longer the only source of astronomical discovery. The Titius–Bode law seemed able to predict the possible existence of as yet unobserved celestial bodies in precise places. The discovery of Neptune was to reinforce this impression.

Logically, the honours for this discovery should have been bestowed on the young English student John Couch Adams, who had recorded his conviction of being able to find that eighth planet in his notebook. But bad luck worked against this modest genius. Bad luck reinforced by the blindness of certain of his colleagues, who, fooled by his youth, didn't believe in his incredible talent. The misfortune of some being the fortune of others, the laurels were to be given to another talented scientist, the Frenchman Urbain Le Verrier, after the affair had almost created a cross-Channel diplomatic incident.

Adams (1819–1892) came from a Cornish farming family. Very early on, he showed exceptional gifts for mathematics. As a child, he often passed his evenings poring over algebra problems. At the age of eight, he overtook his school teacher. At eleven, he could teach a thing or two to the local mathematician. In 1835, Halley's comet gave him his first astronomical shivers. Later, we find him on the benches of the very prestigious Cambridge University, where he stood out for both his scientific excellence and his mild and friendly character. His interest in the unpredictable movements of Uranus came from a report that he had read, signed by the Astronomer Royal, George Biddell Airy, the very person who would soon clip his wings of glory.

If he had not had to finish his studies, Adams' research would no doubt have taken off rapidly. In fact, he had to wait two years before he could return to the problem posed by Uranus and a further two years to solve it. During all this time, the hypothetical planet nagged at him. How far away could it be? He had to have at least a first approximation in order to start his calculations. He used the Titius–Bode law without realising that it was more likely to mislead him than to help him.

His first estimate of the position of Neptune was nearly perfect. It was within two degrees of the truth. However, without observations it was impossible to know this. So Adams set out in search of an observatory that would be willing to verify hs work. He turned to the director of Cambridge Observatory, James Challis, who advised him to submit his work to George Biddell Airy, of the Greenwich Observatory. The latter, as Astronomer Royal, was an influential man, but he wouldn't use his influence for Adams until it was too late.

Airy was not really an unpleasant chap, but he did have two major faults: he preferred the routine to the unexpected, and he believed in the limitations of youth and that nothing matched experience. Moreover, he didn't agree with the scientific choices of the young Adams. The tales of a trans-Uranian planet made him smile. To explain the digressions of Uranus, he preferred his own hypothesis: that the law of gravitation loses its validity the further you get from the Sun. After Saturn, it ceases to act, or almost so, and other forces come into play.

Airy had heard of Adams via Challis. He knew that the Cambridge prodigy was working on the problem but he wasn't interested in the project. To find a planet solely by mathematics was a method that he dismissed out of hand. He didn't believe a few equations could reveal a celestial body.

Adams was probably totally unaware of the reluctance of the Astronomer Royal to help, and he strove to meet him and explain his work. First he gave his notes to Challis. Aware of his student's talents, the Cambridge astronomer could have started observations, but he preferred to rely on Airy's judgment. At the end of September 1845, as he was going home to see his parents, Adams decided to stop off at the Greenwich Observatory to meet its director. Unfortunately, Airy had gone off to a conference in France and wouldn't return for several days. The young man left the introductory letter that Challis had written for him, and went on his way to Cornwall. Airy was later to speak to Challis of his regret of having missed this meeting.

Adams tried his luck again on his way back to Cambridge. This time, the Astronomer Royal had gone to London, but Adams was told that he'd be back in two or three hours at the most. The young mathematician left his card and a summary of his work and promised to come back a bit later. On his return, he was met by a rather unhelpful valet who told him that the director was dining – Airy had the strange habit of having his lunch at three thirty in the afternoon – and that it was out of the question to disturb him. It was probable that the young man's card had not even reached Airy. Adams was puzzled. Was Airy trying to avoid him? Would history have been different if the interview had taken place? At least Adams would have had the chance to defend his work.

Airy eventually read the work of the young Cambridge graduate, and requested him, by post, to send some details. But the questions that the Astronomer Royal asked bore no relation to Adams' approach. What Airy was hoping for was to glean from Adams some comments on his own work, evidence that would support his hypothesis on the weakening of the force of gravity with distance. The letter and its contents left Adams completely perplexed, to the extent that he decided not to respond. Airy was offended. Relations between the two men grew tense.

Meanwhile, on the other side of the Channel, in France, the mysteries of Uranus were stimulating careers. François Arago, the director of the Observatoire de Paris, appreciated for his humility and his habit of encouraging young astronomers to follow new paths, convinced one of his students to consider this problem. Urbain le Verrier, a Norman born in Saint-Lô in 1811, later revealed a character diametrically opposed to that of his teacher: it was to be said of him that even if he wasn't the most detestable man in France, he was certainly the most detested. But at that time he was just a young, brilliant and ambitious researcher. Graduating from the École polytechnique with distinction, he initially trained to be a chemist. He became the assistant of the famous Louis Gay-Lussac, who, despite the excellent qualities of the young man, soon pushed him towards another destiny. This great professor had noticed Le Verrier's impressive talents in mathematics, and when the position of professor of astronomy was created at the École polytechnique, he encouraged him to apply. The Norman got the job. He quickly proved how suited he was to the work.

It was then that Arago met him and advised him to study the thorny question of Mercury. The closest planet to the Sun also posed problems for the astronomers. Its perihelion (the point of its orbit that is closest to the Sun) shifted with time. Most of the shift could be explained by the influence of planets like Venus and the Earth. But there remained 43 little arc seconds that escaped explanation. Le Verrier tried to crack the mystery in vain. It was only much later that Einstein resolved the problem, which is a relativistic effect due to the presence of the enormous mass of the Sun.

The French astronomer next began work on comets, and then Arago drew his attention to Uranus and the possibility that the irregularities in its orbit betrayed the presence of an eighth planet. Le Verrier's method differed from that of Adams, but in the end, the results were nearly identical. The first publication by the Frenchman on the subject appeared 1845 in the *Proceedings of the Parisian Academy of Sciences*. A second followed, a few months later, in June 1846. Airy, who kept himself up-to-date with the latest developments in his science, read Le Verrier's papers and couldn't have failed to compare them with those of Adams. The difference between the two estimates on the position of Uranus was less than an arc second. But instead of reacting to this, the Astronomer Royal only asked Le Verrier a few uninteresting details about his work.

In reply, the French astronomer asked Airy to kindly start a search based on his calculations. The Englishman regretfully declined, since he was about to leave on a journey – five weeks later. While the director of Greenwich persisted in his blindness, his colleagues were less stubborn and urged him to act. Finally, in July, Airy asked Challis to start a search with the Northumberland 30-centimetre telescope. The director of Cambridge informed Adams, who provided him with the up-to-date coordinates. The young prodigy added that the object should appear as a disc of magnitude 9. Despite being given these details, Challis preferred to survey a large region of the sky and to investigate all objects down to magnitude 11.

It was not only in England that people showed little enthusiasm for the search. In France, Le Verrier failed to obtain observing time from the Observatoire de Paris. Tired of fighting, he talked to a young German he knew, Johann Gottfried Galle, of the Observatory of Berlin. Fascinated by the project, the latter went immediately to speak to his director, who agreed to give him observing time. One night was enough, that of 23 September 1845. Galle set himself up in front of the 23-centimetre telescope, one of the best available at that time, and pointed it according to Le Verrier's instructions. While he called out the coordinates of the stars that he saw, his colleague and friend Heinrich Louis d'Arrest found them on a recent stellar map and checked that they matched known objects. Their work had only been going a few minutes when Galle stated another pair of coordinates. But instead of the usual 'On the map!', there was only silence as a response. A few more seconds and d'Arrest exclaimed that the star was not on the map. A few days later, Le Verrier received the congratulations of the Observatory of Berlin and the confirmation that the planet well and truly existed. So now there were eight!

On learning about Le Verrier's discovery, the English were aware that they had narrowly missed glory. They brandished the work of the young Adams, underlining that it pre-dated that of the French astronomer. In Paris, the response was that only official publications counted. This dispute soon went beyond the community of researchers. The French and British press attacked one another in a sad exercise of scientific patriotism. Tempers eventually calmed down thanks to Adams' cool-headedness and humility, when he congratulated Le Verrier on his work and recognised his claim to be the discoverer. The two men even became friends. But while the Englishman led a very quiet life, declining honours – he refused a peerage and the position of Astronomer Royal after Airy – Le Verrier, despite the severe hatreds that his difficult character caused him, took part in a new, very media-targeted search.

Now that Neptune had been sorted out, the problem of Mercury came back to the fire. The new planet had helped to explain the movements of Uranus, would it do the same to those of the closest planet to the Sun? Le Verrier returned to his calculations, brought them up-to-date and compared them with observations. The verdict: Mercury's perihelion was still shifted by 43 arc seconds. The enigma haunted Le Verrier, who in 1859 eventually publicly declared that the Mercurian irregularities were due to the presence of a planet, which he baptised Vulcan.

Some time later, Edmond Modeste Lescarbault, a French amateur astronomer, claimed to have observed on 26 March 1859 the transit of a black point across the Sun. He gave Le Verrier the various coordinates regarding this strange object. Inclination, longitude, eccentricity, transit time, so many parameters that Le Verrier used them to draw up the identity card of this possible intra-Mercurian planet that he called Vulcan: it would have an orbital period of 19 days and 7 hours, a mean distance from the Sun of 0.1427 astronomical units (an astronomical unit equals 150 million kilometres, the distance from the Earth to the Sun), and a mass and diameter much smaller than those of Mercury.

This time, there was a beeline to the telescope. In Zurich, the Solar Spot Data Centre found a few points that could well have corresponded to the object sought. Soon two dozen planetary candidates were lined up. In 1860, a solar eclipse offered the French astronomer the chance to see Vulcan. Since the Moon covered the Sun, it was possible to easily inspect the latter's surroundings without being blinded by its rays. Le Verrier pleaded with all his colleagues to be ready for the event. But Vulcan didn't appear. Little by little, the fever died down. Le Verrier died in 1877 without knowing if his planet was a dream or reality.

FROM PLANET X TO PLUTO

In 1876, a young man born into a distinguished New England family obtained his mathematics degree at the prestigious University of Harvard, near Boston. Percival Lowell was barely twenty, and rather than launching into a scientific career, he went into business. In just six years, he made a considerable fortune. It was then that he changed his lifestyle. He became a travelling writer, and then tried a diplomatic career. The Far East was his favourite region. Indeed he was in Japan when he learned about the sudden blindness of the Italian astronomer Giovanni Schiaparelli, whose work, on the straight lines visible on the surface of Mars, he greatly admired. Lowell decided to continue the work of Schiaparelli. Two factors made his task easier: his mathematical education and his wealth, which in 1894 made it possible for him to finance the construction of an observatory at an altitude of more than 2000 metres, in a remote corner of Arizona, a few kilometres from the town of Flagstaff. The sky there is clear and pure, providing excellent observing conditions.

Percival Lowell didn't remain unknown for long. He vigorously defended his thesis according to which the straight lines are artificial canals constructed by a civilisation in order to bring water from the poles down to Mars' equator. Lowell's story-telling talents, his eloquence and his taste for lectures guaranteed his good reception by the public, but the majority of scientists were deaf to his arguments. He was marginalised. At best, he was considered 'original', at worst, he was considered a crackpot.

To improve his scientific reputation, Lowell began another search, that for a ninth planet beyond Neptune. After all, if the movements of Uranus had revealed the presence of Neptune, those of Neptune might equally lead to a discovery. The only problem was that Neptune was a particularly distant planet. Its maximum distance from the Sun is 4.5 billion kilometres. It was impossible at that distance and with the instruments of the time to hope to detect orbital irregularities. So instead Lowell decided to again use Uranus to find clues to the existence of a trans-Neptunian planet. He busied himself with the calculations while his team at Flagstaff, Carl Lampland and the Slipher brothers, Vesto and Earl, carried out the observations.

The first photographic campaign started in 1905. It finished three years later. Four hundred plates had been exposed and were carefully studied. Every observed region of the sky had been photographed twice, at an interval of a few days, in order to detect the shift of an object, of a possible planet. It was a painstaking task requiring infinite patience. Every image contained many hundreds of thousands of stars. As well as this abundance, there were also false leads, imitations. First of all there are variable stars, the brightness of which changes to the extent that they are sometimes visible, sometimes invisible. There are also comets, which wander without always clearly showing their trailing tails. Finally, there were asteroids, which in certain conditions, behave much like a distant planet would.

The campaign produced nothing conclusive. Percival Lowell decided to try harder. Each time he thought he had better estimates, he sent them to the astronomers, who corrected their observations as a result. Tension mounted when competition revealed itself and threatened to deprive Lowell of his trophy. His greatest adversary was named William Pickering, a graduate like himself of Harvard. In 1908, Pickering announced that after studying the Uranian orbit, it was possible to deduce the presence of a planet which he called '0' of two terrestrial masses, located at about 52 astronomical units with an orbital period of 373 years. In contrast, Lowell's proposed planet, planet 'X', had a mass equal to two fifths of Neptune and was located at some 47.5 astronomical units with an orbital period of 327 years. In other words, the planets proposed by the two men were within a hair's breadth of one another.

In 1911, on the advice of his astronomers – but perhaps also because his adversary had published some new work predicting the existence of three trans-Neptunian planets – Lowell bought a new machine to help with the analysis of the photographs. Called a *blink comparator*, it made it possible to place two plates next to each other and to compare them with the help of a viewer which could be used to look rapidly first at one and then the other. The rapid flickering between one image and the other made it possible to animate the observed stars. If an object moved between the two images, its movement became visible, a true celestial animated cartoon. Would the planet X be animated?

A new three-year photographic campaign started in 1911. It failed like the first. Despairing, Lowell decided to publish his exploratory studies. It was a considerable body of work that he thought would surely interest the scientific community. Alas! no specialised review would accept the paper. Lowell published it using his own funds, but his disappointment was deep. Gradually, he lost interest in planet X. He had nearly forgotten it when a heart attack hit him on the 12 November 1916.

In his will, Lowell left the Observatory a sufficient endowment to enable it to continue its activities without worrying about money problems. But he hadn't factored in the vitriolic reactions of Constance, his widow, who disputed the generosity of her husband to astronomy. The quarrel was only sorted out in 1927, after a long and costly judicial procedure. After the settlement, Vesto Slipher, director of the Observatory, was forced to admit that his financial margin of freedom had been reduced to almost nothing and the Observatory really needed to acquire a new instrument in order to continue its planetary quest. Luckily Abott Lawrence Lowell, Percival's brother, agreed to give him ten thousand dollars to construct a telescope specially designed for the photographic investigation of the sky.

After a few minor technical hiccups, the instrument was assembled and tested in 1929. A twenty-three year old man followed this operation very closely. This was Clyde Tombaugh, who was to operate it. Just a few weeks earlier, he had been pacing the fields of his father's farm in Kansas, forced to accept that a storm had destroyed most of the crops and that he would have to move to the city to find work to support his parents until the next harvest. So his future did not look not terribly promising. Science had fascinated Clyde Tombaugh ever since his father and his uncle had taught him how to look at the stars. Soon, the young boy was making instruments for himself. His favorite telescope was made with bits of agricultural machinery and cars. Not only did Clyde pass hours on end watching the planets, but he drew them in minute details. Wanting to have an expert's comments on his work, he sent sketches to the Lowell Observatory, the only one to his knowledge that worked on planets. The director, Vesto Slipher, was impressed. Just at that moment, he was looking for an apprentice to operate the telescope during the forthcoming photograhic campaign. This was a painstaking task that professional astronomers had neither the time nor necessarily the desire to accomplish, but which would undoubtedly satisfy the young amateur enthusiast. Indeed, Clyde Tombaugh jumped for joy. This job at ninety dollars per month came just when he needed it and saved him from a life of boredom. After a twenty-four hour train journey, he arrived at Flagstaff station early in the afternoon of 15 January 1929. The following day, Vesto Slipher invited him to visit the site. He also took him to the dome which housed the 13-inch telescope. Men were busy finishing the assembly. It was then that Clyde learned that this instrument was destined for the search for planet X. He was on top of the world. What an extraordinary resolution to his problems! He was to closely participate in one of the most passionate scientific searches of the beginning of twentieth century.

Dreams gave way to reality. The telescope was difficult to tame. A manufacturing fault caused many interruptions. A large proportion of the images were unusable. Clyde also encountered the problem of photographic plates that bent in the cold in the dome. The observatory's astronomers considered this a necessary evil, but for Tombaugh, it was a useless evil that had to have a solution. Thanks to him, the number of bent plates soon diminished significantly. The new recruit justified the confidence placed in him.

The third photographic search began in April 1929. It was decided to start by studying the Gemini region: if Lowell was right, it was there that planet X should have been found. Plates were exposed, then subjected to the *blink comparator* a few days later by the Slipher brothers. The latter found nothing. It was necessary to continue and do the round of the constellations of the Zodiac, even if this was a Herculean task. By June, nearly a hundred images had been taken, of which a good part had still to be analysed. Unable to cope, the astronomers asked Tombaugh to also help in comparing the photographs. This meant that Tombaugh was now responsible for all the steps of the research. The task was a big one, but paradoxically this new distribution of work gave him more freedom. He decided, in particular, to modify the observing strategy and to take not two, but three images of each celestial region, in order to increase the chances of discovering the long sought object.

Weeks, then months passed, and there was still no sign of planet X. To make things worse, the telescope would soon approach the plane of the Milky Way, which would increase the number of stars to dizzying heights. Soon there were a million luminous points per image. Finally, in January 1930, the quest came back to its starting point, the Gemini constellation. On 21 January, Clyde Tombaugh photographed a region centred on the star Delta Geminorum. Two nights later, he photographed it again, and then exposed his third plate on 29 January. At that moment, he was like a fisherman whose nets had closed on a fantastic catch, but who didn't yet know it.

The images of Delta Geminorum came to his analysis table on 15 February. The young man placed his eye on the viewer and started the comparison. Three days later, he was still working on it. In the neighbouring offices, the characteristic clicking of the machine was heard. But at 4 pm, the noise suddenly stopped, then started again with a slower rhythm. Clyde Tombaugh didn't believe his eyes. A small fifteenth magnitude point had shifted when he changed from one plate to another. Its apparent movement was that of a planet located well beyond Neptune. He rushed into Carl Lampland's office for him to verify his discovery. Lampland went to check and a few minutes later, he had no choice but to agree. Yes, there was definitely an object in those images that resembled a trans-Neptunian planet.

The team didn't announced the news until 13 March, the day when Percival Lowell would have celebrated his sixty-fifth birthday. The planet was much less massive than had been thought. It took until 1978 and the discovery of its satellite Charon by James Christy to understand that the mass of the planetary couple, located at a mean distance of 39.5 astronomical units from the Sun, was not more than four hundredths of the terrestrial mass.

Having discovered the ninth planet, all that remained was to name it. Vesto Slipher remembered the idea of a young English schoolgirl from Oxford, Venetia Burney, who, inspired by her mythology lessons, proposed calling it Pluto, the name of the Roman god responsible for the kingdom of the dead. This was ideal for the planet which seemed to guard the frontier between the Solar System and interstellar space. Moreover, the initial letters of Pluto, PL, corresponded to initials of Percival Lowell, who had searched for it for so long.

Thanks to his discovery, Clyde Tombaugh became famous. The newspapers were delighted with the story of the farm boy who revealed the existence of a new planet. His adventure was reminiscent of that of the great amateur astronomers like William Herschel. Apart from honours, the young American also won the chance to attend the University of Kansas. He wanted to start at the beginning, but his professors wouldn't allow the discoverer of Pluto take the first-year courses and made him pass directly into the second year. Eventually, he obtained his doctorate in 1939.

Clyde Tombaugh died in 1997, at the age of ninety-one. In the final years of his life, he witnessed the animated debate on the true nature of his planet. Is Pluto really a major planet or is it just a big planetoid? The question became one of hot news in 1992, when the researchers Jane Luu and David Jewitt detected a body (1992QB1) of a decent size (200 kilometres wide) orbiting further out than Neptune. Thanks to this discovery, they confirmed the presence of a disc made up of tens of thousands of asteroids, moving between 30 and 50 astronomical units, in other words at a distance from the Sun of between 4.5 and 7.5 billion kilometres. Named the Kuiper Belt, this disc constitutes a reservoir of short-period comets (the long-period ones come from the Oort Cloud, between 40 000 and 100 000 astronomical units). Pluto and Charon could be two members of this, eminent and massive, but members all the same. Moreover, their constitution, of rock and ice, is further evidence that they belong to the Kuiper Belt.

Certain radicals suggested the declassification of Pluto to that of a minor planet. Others, more consensus-seeking, proposed a double identity: the ninth planet would be regarded as minor to underline its membership in the Kuiper Belt, and as major to take account of the historical importance of its discovery. Defenders of Pluto refused to listen. They insisted that the definition of a planet is a mass sufficient to have to take a spherical form. This is true of Pluto, which has nothing in common with the potato-like asteroids.

In February 1999, the International Astronomical Union declared its intention to put a stop to the debate, which, relayed by the press, had started to take on disproportionate significance. It gave its verdict: there was no question of declassifying Pluto from its rank as a major planet, or at least not before really convincing evidence was found. This evidence might have come from the data transmitted by the American probe Pluto–Kuiper Express, which was intended to be launched in 2006, the anniversary of Clyde Tombaugh's birth. Unfortunately, for budgetary reasons, NASA had to announce in September 2000 that the mission had had to be put aside for the moment.

And planet X? Should it be relegated to the cryptozoological gallery of astronomy? 'Certainly not', reply some, who insist on the fact that nothing can exclude the existence of a tenth planet, or even an eleventh. Nevertheless, these hypotheses remain marginal. The clues are weak and the means of investigation limited. Instruments as powerful as the infrared satellite IRAS didn't find anything. A systematic observing campaign by Charles Kowal at Mount Palomar, which ran from 1977 to 1984, didn't have any success either. But the sky is vast and techniques are fallible. If a significant planet exists beyond Pluto, it reflects so little light from the Sun that it has no problem hiding itself. Moreover, it's possible that its orbit is strongly displaced relative to the plane of the Solar System so that, in contrast with the other planets, it doesn't pass through the constellations of the Zodiac.

In 1999, two astronomers, the American John Matese and the Englishman John Murray announced the hypothesis that there existed a giant planet at about a half light-year from the Sun. Just one of its revolutions around the Sun would last nearly five million years. They deduced its existence from the analysis of the trajectories of long-period comets. According to them, if comets plunge towards the Sun, it's because they've been perturbed by the gravitational influence of a massive body. All that is needed is to wait for the next generation of infrared space telescopes to know if this monster really haunts the outer confines of our Solar System. Paradoxically, it was to be much easier to discover planets around other stars.

4   Why stars wobble

If our stone age ancestors had left us with stellar maps carved into rocks or painted on cave walls, we would have noticed striking differences with today's maps. Several tens of thousands of years ago, the constellations didn't look quite the same. Since despite what people thought up until the eighteenth century, the sky is everchanging. Stars move. They travel. And while this movement is often tiny, or even totally negligible over the scale of a human lifetime, it exists. This is a stroke of luck for astronomers, who found it to be the way to write some of the most beautiful pages of nineteenth and twentieth century astronomy, pages that go by the names of stars like 70 Ophiuchus, 61 Cygnus, Barnard, Epsilon Eridanus or Lalande 21185.

These were the true beginnings of the experimental hunt for exoplanets. The going was tough, with an extraordinary degree of groping in the dark, surprises and failures. In fact, none of the claimed planets of the time were confirmed. Why were there so many setbacks? Probably because the detection methods of the time were stretching limits. A tiny instrumental error was enough to see planets where really there was nothing. Dozens of years went by in a vain scrutiny of the stars in the hope of seeing a possible wobble that would betray the existence of a planet.

THE DISCOVERY OF PROPER MOTION

Although we know him better for the famous comet named after him than for his remarkable scientific accomplishments, it would take pages to list all the achievements (diving bell, wind weather charts, comets, terrestrial magnetism, mathematics, astronomy) of Edmond Halley (1656–1742). This exceptional man was a good star watcher. He trained himself, in particular, by carrying out a southern sky survey in order to enrich the stellar catalogue made by the Astronomer Royal John Flamsteed. Around 1710, he decided to study the maps inherited from Ancient Greece, including that of Hipparcos among others. Patiently, he compared these relics with surveys from his own epoch. Mostly, the maps agreed. However, a few stars disrupted this beautiful unity. Their names are Arcturus, Procyon and Sirius, the sparkling Sirius. Where did the differences come from? The immediate guess was from measurement errors, after all, the instruments of the time were hardly that accurate. Yet, these stars are among the brightest in the sky and so are easily identifiable and measurable.

The other explanation is that these stars had gently slid across the sky over the centuries, and that they were, therefore, animated, moving. In 1500 years, Sirius seemed to have shifted by an angular distance equal to the diameter of the Moon (i.e. about half a degree), while Arcturus had covered about twice that distance during the same time interval. In discovering this stellar movement, which was soon to be called proper motion, Edmond Halley gave the deathblow to the Aristotelian dogma of the stars having fixed positions. With all due respect to the master of the Lyceum, stars wander over the celestial sphere, sometimes quickly, sometimes slowly. But why?

Even though our Galaxy, the Milky Way, is animated by a relatively homogeneous, global, rotational movement, there is some degree of indiscipline. Some stars just do whatever they want. They zoom by, drag their feet, go off at a tangent or escape completely. There are even some that turn in exactly the opposite direction to that of the Milky Way. The reason for these anomalies is the process of galaxy formation, which accumulates accidents, surprises and pot luck.

A galaxy like ours, with its populuation of over a hundred billion stars, was not always like it is today. To reach its present size, it had to indulge in galactic cannibalism. Indeed, we strongly suspect it of having engulfed, during its existence, many small neighbouring galaxies by trapping them in its gravitational net. Has this been enough to satisfy the Milky Way's hunger? Apparently not, since, in a few hundred million years, it's likely to gobble up the Large Magellanic Cloud, a close neighbour located 180 000 light-years away. And before this, it will have snapped up the small galaxy Sagittarius, discovered in 1994 at 60 000 light-years away, as a mere mouthful. The cosmos is a cruel world. It's eat or be eaten.

Thanks to the power of modern telescopes, we now have many magnificent images of incidents of galactic cannibalism. However, despite the enormity of the masses that go into these cosmic contacts, it would be wrong to imagine chain reactions of collisions and explosions. In fact, galaxy mergers take place at a fairly civilised and gentle pace, over several tens of millions of years. And stars in galaxies are so far from each other that in general, even when one galaxy is being swallowed by another, the stars cross paths without even noticing one another. For example, the closest neighbour to the Sun, Proxima Centaurus, is about 4.2 light-years away, or in other words, 40 000 billion kilometres away.

The lack of stellar collisions doesn't mean that things happen in an orderly fashion. There's a good serving of disorder and chaos. All these masses converge and cross over, perturbing each other by their mutual gravity, which is something not to be ignored. Depending on where they've been, stars receive impulses and are shot off in different directions.

However, in order to understand the history of galactic adventures better, it was necessary to work patiently, star by star, surveying as many as possible and establishing the identity of each of them. A formidable task that was carried out by generations of astronomers trained in astrometry, a domain which, as its name indicates, specialises in the measurements of stellar positions, or to be more precise, of stellar positions in the celestial sphere, the apparent sky which can be seen day and night and which is like a cheese cover suspended above our planet. When measuring the positions of stars, astronomy considers this celestial sphere as a curved two-dimensional space, in which two coordinates are enough to place every star. It's similar to the system on the Earth that makes it possible to locate any point by its latitude and longitude. In fact, these ways of measuring the terrestrial globe and the celestial sphere both use angular units, expressed in degrees, minutes and seconds. In this system, a complete loop around the celestial ceiling measures 360 degrees, and from the horizon to the zenith, there are 90 degrees.

In astrometry, these angular measurements are just as useful for determining the apparent sizes of celestial bodies as for determining their apparent movement. So, to the naked eye and from the Earth, our faithful satellite, the Moon, is a half degree in size, or if you prefer, 30 arc minutes. Chance has it that this is also the apparent size of the Sun, which lets us admire its extraordinary corona during total eclipses.

Let's make it clear that astrometry, despite its many talents, cannot do everything. In particular, it cannot detect the radial movements of stars, those movements that are directly along our line of sight. In summary, astrometry is like us. Like astrometry, we can very easily perceive the movement of an object, say, a car, which crosses our path at right angles, but we are no longer sure of anything if the same vehicle zooms straight towards us but is relatively far away. It is moving but nevertheless seems to be still. So, when it comes to detecting radial movements of stars, we need to use another technique, that of radial velocities (we'll get back to this in detail in Chapter 7).

Even in its own domain, astrometry sometimes meets great difficulties. Measuring the apparent diameter of the Moon – which is so big – is relatively easy. In contrast, it is much more difficult to measure small or distant objects, or, worse still, objects that are both small *and* distant. Just think of a star's proper motion. The further the star is from the Sun, the more difficult it is to detect its motion. Experts are often limited to measuring the merest hints of angles. Forget about degrees or even arc minutes, a typical proper motion of a star is about 0.1 arc seconds per year. In other words, such a star takes about 18 000 years to cross a portion of the sky equivalent to the apparent diameter of the Moon. Which goes to show that in order to follow such a movement year after year, it's best to be well equipped.

You should now understand the meticulousness needed to measure the shimmering sky, as well as the need to harvest as many astrometrical data as possible on as many stars as possible, in order to maximise the chances of detecting the various currents that cross the sky. In the Harvard–Smithsonian Center for Astrophysics stellar catalogue of 1966 alone, you can find nearly 260 000 references to proper motions. The data delivered to us by the European satellite Hipparcos on nearly 120 000 stars are much more recent and precise, with an accuracy of the order of a thousandth of an arc second, which corresponds to the height of a man on the Moon seen from the Earth.

While astrometry is seen to be especially useful for studying sets of stars and their dynamics, it is also very convenient for tracing invisible celestial bodies, for being inspired from the most beautiful developments of geniuses like Kepler and Newton, and for juggling with the notions of orbit and the force of gravity.

THE SUN OSCILLATES

Jupiter – and the same is also true for the other planets – goes around the Sun without plunging directly into its flames because although on the one hand it's being pulled inwards by the gravitational force of the Sun, on the other hand it's moving fast enough to escape the pull. A sort of equilibrium occurs, which characterises the orbit.

The force of gravity works both ways. Just as the Sun influences Jupiter, Jupiter influences the Sun. Of course, because of its much lower mass, the planet's power is weak, but it succeeds nevertheless in moving the Sun from its central position and forces it to follow a trajectory around a point that we call the inertial centre of the system, or the barycentre. You can compare the latter to the equilibrium point on a see-saw. When two people of the same weight play together, this point is exactly at the centre of the plank. In contrast, if their weights are different, you have to shift the balance point towards the heavier person in order to give the lighter person enough leverage to be able to lift the heavier one.

The waltz of the stars

The Great Bear
100 000 years ago

All stars move. But seen from the Earth, some move more than others. This is because the path they follow is somewhat different from that of our Sun. Proper motion is this apparent movement of stars on the celestial sphere seen from the Earth.

The task of measuring the positions and the proper motion of stars is that of astrometry. Using astrometry we can follow the distortion over the ages of constellations like the Great Bear.

The Great Bear today

The Great Bear
100 000 years
from now

Complementary techniques

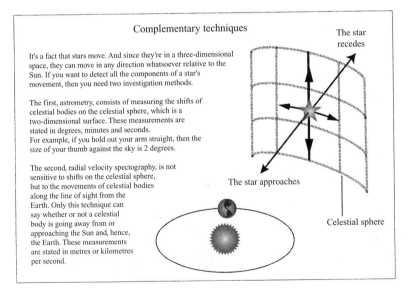

It's a fact that stars move. And since they're in a three-dimensional space, they can move in any direction whatsoever relative to the Sun. If you want to detect all the components of a star's movement, then you need two investigation methods.

The first, astrometry, consists of measuring the shifts of celestial bodies on the celestial sphere, which is a two-dimensional surface. These measurements are stated in degrees, minutes and seconds.
For example, if you hold out your arm straight, then the size of your thumb against the sky is 2 degrees.

The second, radial velocity spectography, is not sensitive to shifts on the celestial sphere, but to the movements of celestial bodies along the line of sight from the Earth. Only this technique can say whether or not a celestial body is going away from or approaching the Sun and, hence, the Earth. These measurements are stated in metres or kilometres per second.

The star recedes

The star approaches

Celestial sphere

Jupiter and the Sun play the same see-saw game, which forces the latter to move in a very small orbit around the barycentre. This perturbation of a celestial body by its satellite has considerable implications for astronomy. It makes it possible to detect planets that are invisible because they're too far away by studying the behaviour of stars around which they orbit.

It's theory that tells us that a perturbation could be a strong clue for the presence of a planet around a star, but could such a perturbation be detected? This is where we see the importance of Edmond Halley's discovery of the proper motion of stars. Thanks to proper motion, we can follow a star for several years and then calculate its path in the sky. If this path is straight, fine. On the other hand, if there is a slight oscillation, we would be justified in asking if something was forcing this star to zig-zag, something like a companion invisible to us, either because it emits too little light (which is the case for certain very faint stars, red or brown dwarfs), or because it doesn't emit any light itself (which is the case for planets).

However, as you've as you probably guessed, the lighter (in mass) the invisible object, the smaller the perturbation induced in the main star and so the less easy it is to detect. If extraterrestrials located on a planet 10 light-years from the Sun observed our star and looked for a planetary perturbation, they would have to attain an astrometric precision better than a milli-arc second (a 3.6 millionth of a degree). And the influence on the Sun of the biggest planet of our system, Jupiter, would be, at this distance, 1.6 milli-arc seconds.

SIRIUS' COMPANION

In 1844, Friedrich Bessel was the first to detect the presence of an invisible companion of a star using an indirect method. The German was a regular in world firsts. In particular, we owe him the first stellar parallax measurement (see Chapter 2), in other words, the first stellar distance measurement. He was undoubtedly a master of precision, which earned him his place of the founder of the German school of experimental astronomy. At the beginning of his career, it was the sea, rather than the sky, that held his attention. He worked at the time as an accountant for a company specialising in marine transport. Being involved with the sea-going world, he quickly learned that it can be useful to know the stars in order to get your bearings across the oceans. During his free time, Bessel studied mathematics and astronomy. His talent was not ignored for long. In 1804, the astronomer Wilhelm Olbers was impressed by the excellence of his calculations on the trajectory of Halley's comet. He recommended the young prodigy to one of his friends, the owner of an observatory near Brême. The ex-accountant attended experimental astronomy classes there and excelled to such a degree that in 1809 he was offered the position of director of the prestigious observatory at Königsberg. He had already developed the habit of extraordinary precision which was to contribute to his fame, always tracking down the slightest sources of error that could ruin the results of an observation.

Glory arrived when, having studied Sirius (he later did the same with Procyon), one of the stars of which Halley had measured the proper motion, Bessel discovered that its celestial trajectory was disturbed by a slight oscillation. There was little doubt as to the explanation: Sirius must be subject to the gravitational influence of an invisible companion orbiting around it. For its epoch, such a conclusion was far from uncontroversial. What had to happen, happened; it shook the entire astronomical community, which hesitated to follow the German on such an audacious path.

Friedrich Bessel died two years later, without having convinced his colleagues of the validity of his calculations. Despite everything, the truth eventually triumphed. In 1862, the researcher Alan Clark achieved consensus by directly identifying Sirius' companion thanks to a powerful telescope. It was not surprising that it was so difficult to detect the object earlier. Its luminosity is particularly faint, ten thousand times less intense than that of the main star. This great difference between the two stars couldn't fail to raise a crucial question. Since Sirius is a close neighbour of the Sun, at only 8.7 light-years away, the luminous faintness seemed to indicate that its companion's mass was small. But if this were the case, then how was it possible to explain how this featherweight succeeded in perturbing the heavyweight Sirius so clearly?

The mystery remained until 1930, the year in which the astrophysicist Subrahmanyan Chandrasekhar outlined the theoretical contours of a very particular celestial body that he called a 'white dwarf'. The Sun and all low-mass stars are destined at the ends of their lives to become white dwarfs. Such stars, after having consumed all their nuclear fuel, throw most of their external layers out into space, revealing a core of degenerate matter of an extremely unusual density, of the order of a tonne per cubic centimetre. A white dwarf can concentrate the mass of the Sun into a sphere as small as the planet Earth. This is why Sirius B, which is both small and massive, is able to perturb its big sister so much without shining brightly.

Detecting the perturbations induced by a stellar mass companion like Sirius B is one thing, but detecting those induced by a planet is another. The higher the companion's mass, the more detectable its influence is. However, a planet like Jupiter is about a thousand times less massive than a white dwarf, so its influence on the main star is also much smaller. Such perturbations were well beyond the limits of nineteenth century astronomers and their instruments.

RUMOURS OF EXOPLANETS

It wasn't until the middle of the twentieth century, with the arrival of more powerful telescopes and the improvement of photographic techniques, that the first attempts at detecting exoplanets started. Several articles about these researches were published in the midst of the Second World War, as if those times of fury and hate stimulated the need to imagine a better world, even if several light-years away. The tone of these articles consciously strayed from the usual scientific rigour, which favours the use of the conditional. Folk were optimistic, like the American–Dane Kaj Aage Strand, an astronomer at the Observatory of Sproul of Philadelphia (Pennsylvania), who in 1943 wrote an article on the star 61 Cygnus. He thought it was accompanied by an invisible object, the nature of which, according to him, was in little doubt as can be seen in his own words:

> The only solution capable of justifying the observed movements shows the presence of a particularly small mass, equivalent to 1/60th of that of the Sun or 16 times that of Jupiter. Since this mass is much smaller than the smallest stellar mass known today, i.e. 1/14th of that of the Sun, the invisible companion must have an intrinsic luminosity so faint that we can consider it as a planet and not as a star. We can thus say that a 'planetary' motion has been detected outside of the Solar System.

For Kaj Aage Strand, a meticulous researcher, to give free rein to so much enthusiasm reflects a certain excitement that had seized the astronomical world at the time. The feat was within grasp, the detection of another world had become possible. It was Dirk Reuyl and Erik Holmberg, astronomers at the McCormick Observatory of the University of Virginia, who had lit the fuse two months before Strand with an article in which they expressed surprise at the strange behaviour of 70 Ophiuchus, a system with two stars orbiting around one another. Celestial mechanics is strict. There are only two ways in which a planet can be found in a double system (also called a binary) without being ejected by gravitational rejection: either it orbits close enough to one of the stars that the influence of the other is negligible, or else it orbits far enough from the stellar couple to treat it as a single mass.

The data accumulated by the two researchers of McCormick Observatory covered nearly ten years. Ten years of regular photographic surveys in order to better measure the celestial wandering of 70 Ophiuchus and to detect a possible perturbation. According to the two American astronomers, there was an invisible companion with a mass of about ten times that of Jupiter in the binary system. Clearly, it was not a star. So, was it perhaps a planet? The authors preferred to avoid the question and said nothing on this point. They chose prudence, but that didn't stop the press from pouncing on the story.

Astrometric detection

In this first example, the astrometric method has detected the apparent motion (the proper motion) of a star on the celestial sphere. After having cleaned up the data, scientists see that the star being studied follows a perfectly straight trajectory. Clearly, it's not subject to any influence, or rather not to any influence detectable by the instrument. We can therefore deduce that that it's a solitary star.

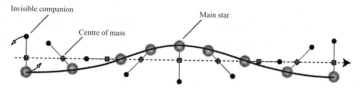

Invisible companion

Centre of mass

Main star

In this second example, the astrometric method has recorded the celestial positions of a star over time. This star generally follows a direction similar to the first star, but it seems to be animated by a sort of oscillation, symbolised by the thick, black line. The cause is a companion that, in orbiting around it, gravitationally perturbs the star and forces it to carry out a small motion around the centre of mass of the system. The oscillation is the result of the successive positions of the star around the centre of mass. Thanks to Newton and Kepler, we can deduce the properties of this curve, the orbital period and the mass of the companion.

Distant planets were a stroke of luck for the press, a gold mine. Unfortunately, the first flaws soon showed up. Strand failed, for example, to confirm the observations of his colleagues from McCormick. From the Observatory of Sproul, he too followed 70 Ophiuchus for several years but disappointingly saw nothing. After the initial fanfare, the quest for exoplanets struck the hard rock of failure.

However, it was only a temporary drawback. What was needed was just to continue to work and to be a bit more patient. Well, Peter Van de Kamp had plenty of patience to spare. It was the necessary companion to his legendary perfectionism. During his whole life, this American of Dutch origin followed the cosmic wanderings of a handful of stars, looking for possible invisible companions, a stubbornness that led to his name being associated with that of a star, Barnard's Star. It was a forty-year scientific marriage, with joys and hopes, with pains and disillusionment.

Barnard's Star owes its name to the person who made it famous, Edward Emerson Barnard (1857–1923). At first, this American hadn't dedicated himself to astronomy. Born in a poor family, he didn't have the means to go to university, so he became an apprentice photographer, an art in which he revealed his excellence and which he used to satisfy his passion for the stars. His astronomical photographs enabled him to detect several new comets and to acquire a certain reputation in the small world of astronomy. In 1883, convinced of his talent, the University of Vanderbilt offered him a scholarship so that he could attend physics lectures. Five years later, Lick Observatory offered him a position as an astronomer. It was then that Barnard detected Jupiter's fifth moon. This discovery guaranteed him international recognition. In 1895, he accepted a job at Yerkes Observatory, at the University of Chicago. Year after year, he confirmed his reputation as a tireless observer, and in 1916, he published an article that caused a huge stir. He described the incredible proper motion of the star that would soon be named after him. With a shift of 10.3 arc seconds per year, Barnard's Star, located 5.9 light-years from the Sun, became the incontestable gold-medal winner in the proper motion competition. It shifted by a lunar diameter every 180 years, which was unheard of. These qualities – large proper motion and close proximity – made it an ideal target for whoever wanted to search for exoplanets by astrometric methods.

VAN DE KAMP, THE UNFORTUNATE PIONEER

Piet Van de Kamp was born on 16 December 1901 in the Netherlands. He was a mischievous child and liked playing tricks and jokes, but he was also friendly and lovable, character traits that he retained during his adult life. He was awarded his doctorate in physics at the University of Utrecht in 1922, and then went to Berkeley in the USA. It was in that country that he conducted most of his remarkable research career, punctuated by numerous prizes and honours. His first work was in the domain of statistical astronomy. As a young researcher he tried to estimate the distance from the Sun to the centre of the Milky Way. He then gradually turned to astrometry in order to hunt double star systems, with a marked preference for systems where the companion is invisible due to its faintness.

When Van de Kamp was named director of Swarthmore College in Philadelphia, the owner of Sproul Observatory, he had already started his research into stellar companions. The new director intensified this research to the point of making it a top priority activity. Normally, the objects dug up by this indirect method are faint, low mass stars, but it's not unreasonable to hope to detect a large planet. And this would be easiest if the difference in mass between the main star and the companion is small and if the possible planet has an orbit quite far, at least several astronomical units, from its sun. If these two conditions were to occur together, then instruments would have the best chance of detecting oscillations.

It happens that Barnard's Star is a red dwarf. Its mass is only a seventh of that of the Sun. If it had a massive planetary companion, it ought to show beautiful perturbations which would be easily visible. It was therefore very natural that right from the beginning, in 1938, Van de Kamp included this star in the stellar sample of his research programme for invisible companions. Nearly every night, at least when the weather was decent, the 61-centimetre refracting telescope of Sproul Observatory was aimed at these candidate stars and they were photographed. Soon images abounded. The meticulous and repetitive work required the measurement of the shifts of the stars in every photograph relative to reference points. After having carried out these measurements, it remained to unravel them, to subtract the movements of the rotation of the Earth in its path around the Sun, leaving only a small residual, if any, that would reveal the presence of a companion. This operation required all the more care given that astrometrists often have to operate at the instrumental limits of their telescopes.

In 1944, six years after the beginning of his programme, Van de Kamp thought he had reliable enough indications of the presence of a companion around Barnard's Star that he talked about it in front of the members of the American Philosophical Society, which met in Philadelphia, the city where he worked. The astronomer showed his first estimate of the mass of the companion: 60 jovian masses.

This was too little to be a star and too great to be a planet. It fell in a sort of middle ground which it's better to avoid explaining, for the moment, except to call it an 'intermediate mass object'. But Van de Kamp wouldn't stop when things were going so well. Every year, he and his colleagues took a hundred new images of Barnard's Star, analysed them and archived them. As data accumulated, the companion's profile became clearer. The oscillations were still there, and were clearer. Nothing less was expected of a man whose works on astrometry and astronomy manuals remain important works of reference.

On 18 April 1963, the American Astronomical Society (AAS) came to Tucson for its annual meeting. Van de Kamp officially announced there in front of his peers that his was convinced that an object orbited Barnard's Star. He acknowledged that the discovery was difficult, that the detected oscillation was a hundred times smaller than the star itself. But his assuredness was now based on more than 2400 images. Since 1944, the companion's mass had melted like snow in the sun. It was no longer 60 Jupiters, but 1.6 Jupiters, while its orbital period was 24 years. In this case, why continue to skirt the issue? It was surely a planet and Van de Kamp said this openly.

His colleagues were more sceptical. They hadn't forgotten the errors of the first detections. Also the planet concerned had a serious handicap that didn't help it to be accepted: its orbit was particularly elliptical. Its minimum distance to its main star, its periastron, was about 270 million kilometres, but its apoastron, its maximum distance from its main star, was just over a billion kilometres. Experts found this surprising when compared with the almost circular orbits of the Solar System planets and they could find nothing in theory to explain it.

The fact that his colleagues were doubtful acted as a spur to Van de Kamp. Five years later, at the annual AAS conference, he presented a new version of his work. He could now claim more than 3000 images of Barnard's Star covering the years 1916–1967 – he also used the photographic archives of the Sproul Observatory. The estimate of the companion's mass was virtually the same as in 1963. In contrast though, the orbit had become even more elliptical. But after all, the Universe is huge and you can well imagine that it allows itself a bit of diversity. So the astronomical community began to come round to Van de Kamp's point of view, and his meticulousness convinced many. Some astronomy manuals started mentioning the existence of an exoplanet.

Nevertheless, Peter Van de Kamp himself felt a little uneasy about the unexpected orbit of his planet. He continued to work on it, and in 1969, he surprised everyone but claiming that there was not an elliptical orbit, but two circular orbits whose combined influence gave the impression of an ellipse. Thus not only did the astronomer remove the thorn that was so painful, but he also introduced a second planet. The object closer to the star went around it in 12 years and its mass was four-fifths that of Jupiter. The other orbited in 26 years and weighed 1.1 Jupiter masses. Everything then fell into place, especially since the two planets orbited in the same plane. Barnard's Star's system started to resemble the Solar System. Could some smaller planets, some earths, for example, be there too?

The couple formed by Van de Kamp and Barnard's Star seemed more robust than ever. After more than thirty years of living together, the passion was still there. This was still the case in 1982 when the researcher wrote a paper in the review *Vistas in Astronomy*, entitled 'The planetary system of Barnard's Star'. It gave, once again, the latest estimates of the mass and orbits of the planets, which were then 0.7 and 0.5 Jupiters, with revolutions of 12 and 20 years. Van de Kamp took advantage of this to remind the reader that exoplanetary astrometry required time and that dozens of years were needed to establish the existence of these distant objects. This, by the way, inspired him to cite the film *Casablanca* to start off his article: '*As time goes by...*' At the age of 44, he had accumulated 4580 images of Barnard's Star, which represented 1200 nights of observation. He was the uncontested champion of Barnard's Star. But this was no guarantee of truth, as is proved by the fact that by beginning of the 1980s, he was possibly the only one to still believe in the existence of his planets.

The first serious doubt was born in June 1973. John Hershey, a colleague of Van de Kamp at Swarthmore College, published an article entitled 'Astrometric analysis of the field of AC + 65°6955 using images from the 61cm Observatory of Sproul refractor'. AC + 65°6955 is a star also known as Gliese 793. Like Barnard's Star, it was one of the stars selected in 1938 for the programme for looking for low mass companions. Hershey had decided to use another measuring instrument than that at Sproul and to submit to it the images of Gliese 793. He chose the automatic machine which the United States Naval Observatory had just bought. What he found left him puzzled, to say the least. In both automatic mode and manual mode, he found a residual oscillation for Gliese 793 similar in every way to that which animated Barnard's Star.

Only one of two possible explanations could be right: either Gliese 793 had a planetary system perfectly similar to that of Barnard's star, or else the perturbation was an illusion induced by the Sproul telescope. Could it be that Van de Kamp had been fooled by an instrumental problem? Everything pointed that way. In any case, Hershey reported these troubling hints. His analysis showed that Gliese 793 had been affected by two hiccups; one, very strong, in 1949, the other, weaker, in 1957. And it was on those two dates that the Sproul telescope had been undergoing maintenance. In 1949, a new cell, a sort of ring which holds the objective lenses, had been installed. If the ring were too tight, it could deform the lens and give a different sky image to the preceding one. If it is too loose, the lenses may move during telescope movements. And in 1957, the objective lens had been replaced.

Even though it wasn't its aim, this study struck a serious blow to Van de Kamp's planetary hypothesis. And if this first doubt was not enough, a second followed shortly after. This came from a young researcher from the Allegheny Observatory (University of Pittsburgh), George Gatewood, advised by his mentor, Heinrich Eichhorn, of the University of Southern Florida. The two men based their study on 241 images of Barnard's Star taken with the Allegheny telescope between

1916 and 1971. This work put the nails in the coffin of Van de Kamp's planets.

CLASSICAL ASTROMETRY FAILS

Ironically, George Gatewood had already met Peter Van de Kamp in 1966, during a reception at a hotel in Florida. He was then only a young recently graduated student, but he had the courage all the same to approach the great astronomer to confide in him how much he admired his work. He also told him of his intention to specialise in the field of astrometry. However, it was not, at the time, the quest for exoplanets that stimulated Gatewood's interest, he dreamt of a comparative study of American telescopes involved in astrometric investigation of the sky. In this way he would be able to determine the small instrumental errors of each one, which make it difficult to use photographs from different origins for any single study. Ideally, such a study would make it possible to create a standard and reliable database, accessible to everybody.

George Gatewood left the University of Florida for the University of Pittsburgh with the clear intention of carrying out this instrumental comparison. Events decided otherwise. One day, the director of the Observatory showed him a study that contradicted the work of Van de Kamp on Barnard's Star. The author, Nicolas Wagman, maybe from shyness, refused to publish the results. It was suggested it would be a good idea if someone took up the baton, carrying out a new sequence of measurements and publishing the conclusions. Gatewood didn't have the least desire to commit himself to such a task. Neither did Eichhorn, his thesis supervisor. Neither of them had any doubts regarding the existence of Van de Kamp's planets. Why redo the work? Van de Kamp had collected hundreds of photographs and his reputation as a meticulous researcher was unassailable. So, the first time he was asked, the young Gatewood refused to take up this thesis subject. However, the director asked again a few weeks later. He was convinced of the value of the research. And it was no longer a proposal that he made to the young student, but an order. Take it or leave it.

Gatewood and Eichhorn decided to respond to ill fate with good science. They knew they had fewer images than their colleague, but they overcame this drawback by choosing eleven reference stars – against Van de Kamp's three – in order to measure the shifts of Barnard's Star as precisely as possible. Moreover, Eichhorn, a great lover of mathematical techniques, invented a new way to reduce the data of stellar proper motion. After several months of work, the two men were ready to deliver their conclusions: there were no planets around Barnard's Star, at least none that resembled those described by Van de Kamp. This thesis was the theme of an article published in 1973 in the *Astronomical Journal*. Its title leaves little room for doubt regarding the results: 'An unsuccessful search for a planetary companion around Barnard's Star (BD + 4°3561)'.

Thanks to his thesis, George Gatewood became a leading figure in astrometry. Gradually, against his wishes, he gained a reputation as a gravedigger for planets. Not one survived his unequalled counter-tests. The candidates around Lalande 21185, 61 Cygnus, Barnard's Star and Alpha Centaurus went into the dustbin.

Meanwhile, Van de Kamp continued on his chosen course. In 1974, he published an article on another star, Epsilon Eridanus, which is quite similar to the Sun but a bit less massive. Like Barnard's Star, it was part of the original stellar sample from 1938. Van de Kamp and his collaborators had photographed it no less than 900 times over forty years. After a meticulous study of the data derived from these photographs, the American–Dutch astronomer concluded that there was a planetary mass equivalent to 6 Jupiters orbiting Epsilon Eridanus in 25 years. However, the conclusions of this study were also invalidated by later work by other teams. It was time for astronomy manuals to talk of exoplanets in the conditional once again.

Despite these negative opinions, Van de Kamp refused to accept what became more and more difficult to deny. He continued his crusade against the infidels. In 1977, he concluded an article with a citation taken from St John's Gospel (20,29): 'Jesus said: since you saw me, you believed. Blessed are those who have not seen but have believed!' Unfortunately for Van de Kamp, in science, it's less a question of faith than of tangible evidence. This great researcher, universally recognised for everything that he brought to astrometry, died on 18 May 1995. According to George Gatewood, Peter Van de Kamp believed in the existence of his planets right to the end. A little before his death, the two men had met at a private reception and Van de Kamp had continued to support his thesis. It was also at this moment that he advised Gatewood to stop looking for systematic errors in other people's work and to finally take the risk of making some himself. It was his way of suggesting to his colleague that it was time for him to take his turn in the search for exoplanets. The advice did not go unheeded.

A year after Van de Kamp's demise, during the annual conference of the American Astronomical Society in June 1996 in Madison, Wisconsin, Gatewood announced that he had detected a planet around Lalande 21185. Located at 8.2 light-years from the Sun, this red dwarf was animated by a relatively large proper motion of 4.8 arc seconds per year. It was therefore an ideal target for the astrometrical detection of a low mass companion. It had already been the subject of an article by Sarah Lippincott in 1960, then a collaborator of Van de Kamp, in which she had shown that the remainders of Lalande 21185's proper motion apparently hid the signature of an intermediate mass companion characterised by an orbital period of about 8 years. However, in 1974, a young astronomer by the name of George Gatewood had swept the planet away with the brush of his counter-tests. He even supported this first judgment with a second study in 1992, for which he had used a new measuring instrument, the extreme precision of which was ensured by its use of a laser. No, he confirmed at the time, there was no planet around that star. And yet...

However, after spending so much time examining Lalande 21185 and observing it in all sorts of circumstances, Gatewood ended up detecting a perturbation, which appeared to be due to not one but two planets. The first would have a mass slightly less than that of Jupiter and an orbital period of 5.8 years. The second, more massive, would weigh 2 Jupiters and complete its orbit in about 35 years. For three months, Gatewood had tried everything to invalidate these measurements. He had no desire to go through the experience that he'd subjected other astronomers to, that of seeing his candidate planet reduced to a figment of the imagination. However, when he thought he had got rid of all possible sources of error, he learned about the discovery – and the confirmation – of the exoplanet around the star 51 Pegasus by a method other than astrometry. The mood was festive. The astronomical world regained confidence. Gatewood joined in the celebrations by orally announcing the discovery of planets around Lalande 21185 during a conference. He hoped that another team would confirm his work.

Despite the efforts of the American researcher, it's unlikely that classical astrometry, as carried out by Van de Kamp or Gatewood, could ever have succeeded in carrying off such an exoplanetary trophy. In fact, this detection method has been superceded, in the domain of exoplanets, by that of radial velocities. Given the competition, financial resources started to decline. In 1999, George Gatewood continued his planetary investigations on an essentially volunteer basis. American science is tough on its researchers and requires rapid results from them. There's now little chance that the 100 000 images lying dormant in the archives of Allegheny Observatory will be used for the quest for exoplanets, except every now and then on special occasions. The future belongs to the new astrometry, which can rely on spatial telescopes and on interferometry (a technique combining light from several mirrors) to achieve real feats.

So the experts in classical astrometry have not had the pleasure of discovering the first exoplanet. Undoubtedly with great reluctance, they had to leave this privilege to other researchers, who didn't at all expect to make such a discovery. And the whole astronomical community has still not recovered from the surprise of having witnessed such an event.

5   Neutron planets

The Universe is a zoo inhabited by exotic creatures. The celestial menagerie reveals the creativity of physical forces: forces that astrophysicists untiringly try to explain by theory, experiment and observation.

In the 1930s, researchers like Lev Landau, Robert Oppenheimer, George Volkoff, Fritz Zwicky and Walter Baade, having gone through the calculations, became convinced of the theoretical existence of a star never hitherto observed. It was an extremely dense star, the core of which was just an aggregate of neutrons. Does it really exist? This question was asked for nearly thirty years until thanks to the observations of a young Irishwoman it was possible to confirm the theory. But what that theory could never have predicted was that one day a Polish researcher, employed by an American university and working on a radio telescope in Puerto Rico, would discover the first exoplanets around one of these dizzying stars.

PULSAR, YOU SAID A PULSAR?

At the age of eleven Susan Jocelyn Bell failed the entrance exam that would have enabled her to attend a state grammar school. However, her father, an architect who was curious about everything and astronomy in particular, instead sent her to a private school, where she thrived. Perhaps she owed her success there to her physics teacher, whose enthusiasm for the subject was matched by an ability to explain it. Whatever the reason, Susan developed a passion for her chosen subject that was to lead to one of the major astronomical discoveries of the twentieth century.

After obtaining a degree in physics at the University of Glasgow, this young woman worked at the prestigious Cambridge University with Anthony Hewish and his team at Mullard Radio Astronomy Observatory. The adventure promised to be an exciting one. The observatory was putting the finishing touches to the construction of an imposing radio telescope. Radio astronomy was then a very young discipline, the heir of the radar techniques invented during the Second World War. As its name indicates, radio astronomy uses radio waves. It can detect objects and events that are manifested in regions of the electromagnetic spectrum (the spectrum which groups together many sorts of radiation: radio, infrared, visible, X, gamma, etc.) to which our eyes are not sensitive. The Cambridge radio telescope was expected to be able to study a cosmos very different from that seen by optical telescopes, a cosmos full of vibrations, oscillations, all sorts of sounds. The leader of the project, Anthony Hewish, particularly wanted to observe quasars, the hearts of galaxies which emit a continuous flow of radio signals and which are suspected of harbouring black holes as massive as 100 million suns.

Obviously, there were some drawbacks intrinsic to this fledgling radio astronomy. Those who used it were condemned to ploughing through kilometres of printed paper containing the rather indigestible translations of what the telescope had heard. In Hewish's group, no-one escaped this meticulous work, Jocelyn Bell included. In 1967, when she was only twenty-four, she took up her share of the deciphering. Looking through the reams of paper, she noticed a weak signal, which repeated with an emphatic persistence every 1.3 seconds. The student alerted her research supervisor. The team was puzzled. In any case, the signal wasn't emitted by a quasar. It was too rapid. Its perfect periodicity pointed to a human origin, possibly a satellite or perhaps television broadcasts or radar emissions? One after the other, all hypotheses were analysed in detail. None was retained. Lacking further ideas, the Cambridge researchers ended up calling their object LGM1, an abbreviation of 'Little Green Men 1'. Were little green men sending us messages via cosmic waves? The hypothesis was crazy, but in the absence of any consistent explanation...

It didn't take long for the scientific community to spoil the extraterrestrial hypothesis. The discovery a little later of a second fast periodic source, located on the other side of the sky (LGM2), made it unlikely that the origin was artificial. Hundreds of physicists pondered upon the mystery. Some then remembered the theories about neutron stars developed in the 1930s. The Italian Franco Pacini suggested that some of these stars had to look like very concentrated energy sources. Enveloped by an intense magnetic field and subject to a rapid rotational movement, like spinning tops, they would generate periodical puffs of radio waves, and Pacini proposed that it was these that had sometimes been caught by radio telescopes. All that remained was to name the phenomenon: it was called a 'pulsar', an abbreviation of 'puls(ating) (st)ar'.

In 1969, a pulsar, PSR 0531+21, was discovered in the heart of the Crab Nebula, in the constellation of Taurus. This corner of the sky is well known to astronomers, and has been for centuries. It was here that in 1054, Chinese astronomers noticed and listed in their records the appearance of a new star, so bright that at first it was possible to see it in broad daylight. They had witnessed one of the most cataclysmic events in the Universe: the explosion of a supernova, a giant star that has reached the end of its life. The discovery of the Crab pulsar, 5000 light-years away, allowed contemporary astronomers to establish the intimate link between supernovae and neutron stars.

It remained for astrophysicists to reveal the intimate mysteries of a pulsar, starting with its birth, which, paradoxically, starts with the death of a star. In short, to understand a pulsar, you need first to understand stars, how they form, how they live and how they die.

THE AGITATED LIVES OF STARS

At first glance, the thousands of bright points that dot the night sky all look the same. If you try hard, you can distinguish a few differences in brightness and size. But if you continue to look up for a few minutes, you soon notice that not all the stars are the same colour. There are some that are white, of course, but also there are blue ones, red ones, yellow ones, orange ones. This multi-colour palette gives a first clue to the extraordinary variety of types of stars. Every one has its own identity: old or young, big or small.

A star always begins to form in the same way. An interstellar cloud of very sparse primordial hydrogen calmly floats in the Cosmos until an event – it could be, for example, the crossing of the galaxy by a density wave – creates within it some zones of higher density. Molecules of gas stick together. They then profit from their newly accumulated mass to pull in more gas around them. The force of gravity does its work. The more the bubbles of gas grow, the higher the pulling power given to them by gravity. Soon, there's nothing but a big gaseous blob. From a few thousand particles per cubic centimetre, the density quickly passes to billions of particles per cubic centimetre. The protostar is about to light up. It just has to attain the right conditions: an internal temperature of 10 million degrees and a pressure of a billion atmospheres. The star has no problem attaining these levels if it has the critical mass.

The force of gravity continues its task. Each layer of the star, attracted by the higher density core, weighs down on the layer below. The star is compressed and compressed yet again. The more the vice is squeezed, the more the atoms are agitated as they speed up. At around 10 000 degrees, the hydrogen atoms lose their electrons and become positively charged ions, or in other words, protons. At that instant, it's no longer an ideal gas that forms the star but plasma. The gravitational onslaught continues. The hydrogen ions continue to move ever more vigorously in all directions. And the temperature keeps increasing. At 10 million degrees, the electromagnetic force of repulsion which forces two particles of the same charge to stay apart is overcome. The hydrogen ions reach such a speed that some of them succeed in fusing together, releasing a strong burst of energy. The thermonuclear process has just lit up in the heart of the star.

It's then that gravitation first meets a worthwhile adversary: the force of gas pressure. The hotter a gas is, the more it expands. Just think of a pressure cooker sitting on the stove to see this (the lid being the equivalent of gravitation in this case). In a star, the gas doesn't stop expanding during the gradual increase in the temperature; but with the start of thermonuclear reactions, the force of the gas pressure becomes capable of matching the force of gravitation. Hydrostatic equilibrium is then attained.

Theoretically, all stars that have a mass above 0.08 solar masses succeed in igniting hydrogen fusion reactions, which at the end of a complex process, produce helium nuclei. Our Sun has been consuming hydrogen for about 5 billion years. And it's likely that it will continue to do so for another 5 billion years. It owes this great longevity to its small mass. In contrast, other much more massive stars consume their hydrogen in just a few tens of millions of years. This is because at the heart of these stars everything reaches impressive levels, whether it's gravitation, temperature or gas pressure. This leads to nuclear reactions that take place at a much faster rate than in light stars and the fuel is used up much more quickly.

But whether it's massive or not, and whether its life is short or not, a star passes through different stages which characterise its evolution. Each of these stages is characterised by the nuclear fusion of a new dominant element. As we saw, a stellar birth always starts with hydrogen fusion. A star stays in this state for most of its life, but there comes a day when the initial fuel is used up. Does this mean the death of the star? Not yet. Hydrogen fusion, as well as producing energy, has also made helium. It's helium which then becomes the dominant element in the stellar core, and which nourishes the thermonuclear reactions.

When the first cycle of fusion has finished, the temperature of the core of the star drops suddenly, and with it the gas pressure. Gravitation can then start a new attack. But in doing this, it provokes an excess of movement among the helium nuclei, or in other words, a new rise in the temperature. In turn, the helium nuclei become agitated enough – at a temperature well above that which was needed for hydrogen – for them to overcome the force of repulsion. The fusion process restarts, this time with helium. With the help of thermonuclear reactions, the plasma dilates until it creates a new hydrostatic equilibrium. But soon the helium is used up in turn. There's another drop in temperature, a new weakening of the dilation force and a new attack by gravitation, until the core of the star attains the conditions required for the fusion of a new dominant element, carbon.

Heavyweight stars pass through a larger number of steps because their enormous mass allows them to go further than the stage of carbon fusion. Their cores soon attain a billion degrees, the temperature that characterises the beginning of oxygen fusion. Then it is silicon's turn and finally iron. The temperature of the core is then close to 5 billion degrees, which is high enough to continue thermonuclear reactions. However, since iron is the element with the highest binding energy, after this, fusion of heavier elements uses up more energy from the star than it provides. So the party is spoilt. The temperature drops, the gas pressure drops too, and gravitation goes crazy. The star is ripped apart: it becomes a supernova, an example of one of the most cataclysmic phenomena known, capable of emitting more energy than 100 million suns.

While the outer layers are ejected into space at speeds of the order of 30 000 kilometres per second, the core implodes and collapses. If its mass exceeds one and a half times that of the Sun, it becomes a black hole, an object so extraordinarily dense that even light can't escape from it. If it's below this limit (called the Chandrasekhar limit), it becomes a neutron star, the density of which is comparable to that of an atomic nucleus. This time, it's not dilation of the plasma that opposes gravity, but neutrons which flee from each other in order to respect a strict law of the quantum world (called the Pauli exclusion principle) which forbids them from being too close to one another. This agitation means that a neutron star is incompressible beyond a certain limit. The matter that constitutes it degenerates into a state that is ruled only by the strange laws of quantum physics.

This stellar transformation is not at all commonplace. In just a few instants, a mass equivalent to that of our Sun is gathered into a sphere with diameter of barely 20 kilometres. Just like a skater

Stars die too

The destiny of solar mass stars

Initial size of the star

At this stage, the star is fed by the thermonuclear reactions of its most abundant element, hydrogen.

Giant

After having burnt all its hydrogen, the star goes on to helium, then carbon. It can't go further and transforms itself into a planetary nebula.

Planetary nebula

In the final stage of its existence, a giant star throws off its outer layers, leaving a white dwarf at the centre.

White dwarf

A white dwarf is a dense star that concentrates a mass equal to that ot the Sun into a volume equal to that of the Earth.

The destiny of massive stars

Initial size of the star

At this stage, the star is fed by the thermonuclear reactions of its most abundant element, hydrogen.

Supergiant

A massive star first burns hydrogen, then helium, carbon, oxygen, silicon. Finally, it hits the ceiling with iron.

Supernova

A supernova is an explosive event that throws out most of the matter from a star.

Black hole or neutron star (pulsar)

The core of a supernova yields, in collapsing, a black hole or a neutron star. In the latter case, the star rotates very quickly and emits radio waves. It's a pulsar.

who accelerates his spin by bringing his arms in to his body; a star that passes from a considerable size to a tiny diameter experiences a dizzying acceleration which pulls it into a spinning top. This crazy rotation, together with the existence of a considerable magnetic field, contributes to pulling electrons from the surface of the star and accelerating them along field lines at speeds approaching the speed of light until they yell out a 'shriek of light' called synchrotron radiation. This is the mysterious radio source received by radio telescopes. And if it's periodic, it's because the pulsar's beams – there should be two, one at each pole – behave like light from a lighthouse and sweep the Earth as the neutron star rotates.

Having studied them, experts realised the extraordinary regularity of pulsars' impulses. They're able to compete with the regularity of atomic clocks, true chronometers to make the best Swiss watchmakers green with envy. This quality turned out to be critical for the next episode of the adventure.

RUMOURS ABOUT PLANETS

In 1970, just one year after the discovery of the Crab pulsar, David Richards, an American astronomer who had followed the object for several months and minutely timed the arrival times of its radio flashes, announced that he had detected an anomaly in the periodicity of the signal. There were only three possible explanations for the origin of this irregularity. It could have been an effect of precession, that is, an oscillation induced by the star itself, the consequence of vibrations in the star due to structural instabilities, or finally, a perturbation induced by a planet of which the orbital period was close to 11 days.

The third hypothesis made many smile. How could a planet exist around a star as exotic as a neutron star? The sceptics were right because a new study soon concluded that the anomaly in the Crab pulsar was due to an internal instability of the star. But a new alert occurred in 1979 when a Polish team led by Marek Demianski proposed the possible presence of a planet around the pulsar PSR B0329+54, located in the constellation Camelopardalis. The proposed planet would have a mass just over half that of Earth and it would complete a revolution in 3 years. Here again, the three hypotheses were considered, and here again, that of internal perturbation of the star won.

Despite these failures, some researchers investigated the subject and asked themselves if it was theoretically possible to find planets around a pulsar. Was it possible for planets to survive the supernova? The answer is very probably no. You sometimes read that a supernova pulverises every body close to it. In reality, ejection is a bigger problem than destruction. In exploding, a star loses a substantial part of its mass. Its gravitational influence suddenly falls. Its furthest planets, freed from this force of attraction, fly off in straight lines, towards the cosmos. Others, in lower orbits, probably no longer exist. In fact, as the star goes through its different stages of fusion it increases in volume. Its outer layers puff out enough to enable it to reach the giant stage and possibly even that of supergiant. When the Sun has finished burning its hydrogen, it too will experience expansionary phases that will engulf the Earth. And when a planet finds itself taken into this mass of stellar gas, its circular parth is perturbed and converted into a spiral that inevitably leads to the core of the star. All of which means that for a planet to survive a supernova would be a miracle.

Following the disappointment of 1979, nothing more was heard about pulsar planets for 10 years, but interest in the field was not dead, merely dormant. In 1991, it was the turn of Andrew Lyne, a reputable British researcher, to announce a planet around a pulsar – this time it was PSR 1829–10, which is located in Sobieski's Shield (Scutum), 30 000 light-years away. It was the start of a scientific adventure full of upsets.

In 1967, when news of the discovery of the first pulsar hit the world, Andrew Lyne, a student, was working on his thesis on the theme of lunar occultation at Jodrell Bank, UK. Lacking a sophisticated telescope, scientists then used the Moon to better read the radio sky. In this method, the radio telescope was pointed at the radiosource as the Moon passed in front of it. By studying the way in which the signal disappears or reappears, you learn more about its properties. Even if the procedure nowadays seems antiquated, at the time it made it

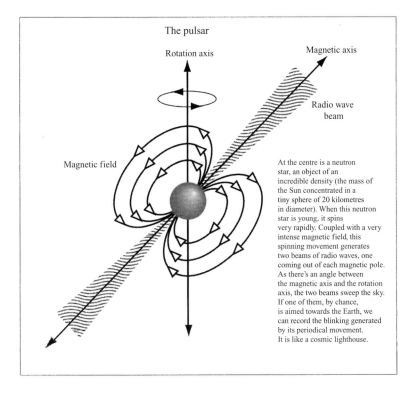

The pulsar

Rotation axis

Magnetic axis

Radio wave beam

Magnetic field

At the centre is a neutron star, an object of an incredible density (the mass of the Sun concentrated in a tiny sphere of 20 kilometres in diameter). When this neutron star is young, it spins very rapidly. Coupled with a very intense magnetic field, this spinning movement generates two beams of radio waves, one coming out of each magnetic pole. As there's an angle between the magnetic axis and the rotation axis, the two beams sweep the sky. If one of them, by chance, is aimed towards the Earth, we can record the blinking generated by its periodical movement. It is like a cosmic lighthouse.

possible to record data at high speed and to study rapid and periodic phenomena of the cosmos.

When the young Lyne heard about his British colleagues' pulsar, he instantly decided to put his thesis aside and to start hunting for neutron stars. The Jodrell Bank radio telescope, 76 metres wide, was up to the challenge. Using it, the British team were able to produce the second paper in the history of the search for pulsars, a few weeks after that of Jocelyn Bell. A year passed, entirely dedicated to this search, then Andrew Lyne went back to his thesis, but once it was finished he yielded again to the call of the pulsars. He dreamed of studying them not for themselves, but for the science that could be done with them. The pulsar became an extraordinary instrument of scientific investigation, for verifying, for example, certain effects predicted by relativity.

In 1993, John Taylor and Russel Hulse, two researchers from Princeton, were honoured with the Nobel Prize for their work on

PSR 1913+16. This pulsar, discoverd in 1974, had the idiosyncrasy of belonging to a binary system of which the second member was another pulsar. With two such massive and close objects (only 1.8 million kilometres separated them), the Americans told themselves that it had to be possible to measure relativistic effects. So Taylor and Hulse timed the radio impulses of PSR 1913+16 for several years. Finally, they succeeded in finding evidence for anomalies that showed that the two pulsars were approaching one another by several metres per year. There was only one plausible explanation: the hypermassive couple emits gravitational waves that perturb the orbits and force them to approach one another. In 300 million years, the two titans will merge.

At Jodrell Bank, Andrew Lyne and his colleagues continued their search, with patience and perseverance, and found dozens of pulsars. PSR 1829–10 showed itself for the first time in 1985. As they confirmed and refined their measurements, the British realised that there was an anomaly in the arrival time of the impulses from PSR 1829–10. Months passed, and the irregularity persisted. The team tried to work out its origin. Every hypothesis was laid out for consideration, but after much examination, it was the most troubling hypothesis that best explained the observations. The perturbation seemed to be induced by a companion orbiting around PSR 1829–10, a planet whose mass must be less than that of Uranus, while its orbital period was close to 6 months.

This last detail was what most bothered the Britons. Six months happens to be half a terrestrial year. Could it be that the variation in the radio impulses was due not to a movement of the pulsar with respect to the Earth, but to a movement of the Earth with respect to the pulsar? The confusion was possible, if one wasn't careful. All it required was to make an error in data reduction to fall in the trap.

We've already pointed out that the radio flashes arrive here with an exceptional regularity. In the case of the most stable pulsars, the time delay that separates the two radio emissions only changes by a billionth of a second every thousand years. Only atomic clocks are accurate enough to use to time pulsars to give a sufficiently detailed profile of their behaviour and detect the possible presence of a companion.

When a planet goes around a neutron star, the latter, as we've already seen, is slightly shifted from the inertial centre of the system and goes through a circular movement, which has consequences for the arrival time of the radio impulses on the Earth. As the time differences are tiny fractions of a second, these offsets are not detectable to start with. But by accumulating them, they can soon be revealed by meticulous work.

As in the case of the astrometric method based on parallax, data resulting from the measurement of a pulsar must be corrected absolutely for their dependence on the rotation of the Earth itself and its annual tour around the Sun. Computers are the ideal tool for carrying out the calculations for these corrections. The programme used by Andrew Lyne when he made his discovery assumed that the Earth has a circular orbit. In fact, the Earth's orbit is very slightly elliptical, but the difference is not significant as long as the position of the pulsar is determined precisely. But such a precision is only possible by accumulating observations of the pulsar. This is almost an automatic procedure. But no-one is immune to a slip in concentration. At the time, the Jodrell Bank team was juggling with several pulsars whose positions had to be determined with an error smaller than a fraction of an arc second. By a stroke of bad luck, the position of PSR 1829–10 had not been corrected. The margin of error on its position was greater than 7 arc minutes. This was to have disastrous consequences.

After months of hesitation, Andrew Lyne and his colleagues Matthew Bailes and Setnam Shemar decided to publish their discovery. The article came out in the review *Nature* on the 25 July 1991. Very quickly, experts worried about the six month orbital period. But they knew the meticulousness of Lyne and finally allowed themselves to be convinced. Titles announcing the first exoplanet flourished.

ONE PLANET LOST, TWO FOUND

At the end of 1991, in the lull between Christmas and the New Year, Andrew Lyne took advantage of the peace and quiet to get back to a bit of research. Once more, he looked at the case of the pulsar PSR 1829–10. Suddenly, a thought sent shivers down his spine. Was it possible that the 6-month period was really an artefact of the Earth's rotation about the Sun and that the anomaly he thought he'd found with PSR 1829 − 10 was due to an error in calculating the celestial position of the pulsar, which was subsequently amplified by the data correction program. The Briton immediately went back to all the data he had available, recalculated the position of the star and introduced the new parameters into the programme. A few seconds later, the computer returned its verdict. There was no anomaly. The planet had vanished into the computer circuits. Lyne remained still for several minutes, struck dumb, facing the screen that displayed the terrible result. He couldn't understand how he and his team could have let by such a stupid error. He had to confess it as quickly as possible. He decided to go to a conference organised in January in Atlanta and to contact the journal *Nature* to request the publication of an erratum after his presentation.

On 15 January 1992, Andrew Lyne climbed onto the podium and started his talk. Apart from a few murmurs, there was silence in the room, a disappointed silence. The Briton concluded by apologising profusely to the whole scientific community. His words had barely been pronounced when a thunder of applause rang out. Clearly, the audience appreciated for its true worth the courage of their unfortunate colleague.

While Andrew Lyne was leaving the stage, the auditorium was still humming with the surprise news. His audience barely noticed that another speaker had taken his place and was ready to speak. He was one of the few who had heard the news the day before. Andrew Lyne in person had warned him in order that he wouldn't be too upset by the news and so risk being taken aback during his presentation. The new speaker was a Pole called Alexander Wolszczan and he was about to announce the discovery of two or maybe even three planets around the pulsar PSR 1257+12.

In 1952, when he was only six, the young Alexander Wolszczan was running around the Polish countryside near his house when, foolhardy and reckless like most children of his age, he decided to try to cross a barbed wire fence in a single leap. The failed attempt left him with a chunk cut out of his knee and forced him to stay put for several days. He was bored stiff. Right up until the evening when his father, a professor of economy and an astronomy lover, took him on his shoulders to see the celestial ceiling. He taught his son about the constellations. The virus was transmitted. The young Alexander soon plunged into astronomy books.

When he arrived at the threshold of higher studies, he quite naturally chose physics. He enrolled at the University of Torun, which still resounds with the feats of Copernicus. Due to tradition, astronomy holds a privileged place there. The only drawback was that the faculty suffered from a lack of equipment. Luckily for the researchers, communist Poland was not as closed as its partners in the Eastern bloc and it wasn't too hard to obtain temporary visas to visit foreign laboratories. So the student Wolszczan went several times to the Max-Planck Institute in Bonn, which has a radio astronomy department.

At first, Wolszczan aimed at optical astronomy, but after having noticed that the field was already chockablock, he shifted his sights to radio astronomy, which was much more promising. It was the time when pulsars were hitting the headlines. In West Germany, the Polish student honed his skills on the Effelsberg radio telescope, which, with its 100-metre diameter, was then the largest orientable dish in the world.

Life was kind to the young Wolszczan. But everything changed in December 1981 when General Jaruzelski, whose dictatorship directly opposed the democratic resistance of the Solidarity union, came to power. Frontiers were sealed. Wolszczan, who was about to leave for another visit to Bonn, had his visa refused, the first time that this had happened. He tried and tried again. After a few weeks, he finally obtained a travel authorisation for West Germany and left with the definite intention of not returning to his country as long as the military regime imposed fear and oppression. It was a painful decision. His wife and daughter couldn't follow him. He had no idea when he would see them again, even though he intended to move heaven and earth in order to enable them join him. Finally, after several months of unbearable waiting, visas were delivered and the little family was reunited.

The Wolszczans lived in Bonn for just over a year, until one day Alexander responded to a job offer from the American Cornell University, which was looking for a resident radio astronomer for its Arecibo telescope, in Puerto Rico. With its 305-metre diameter, nestled in a natural dip in the island, this dish is the largest in the world. Only its secondary antenna, suspended at a height of 130 metres, is steerable and can be used to look at the sky up to 20 degrees from the zenith (from directly overhead). The change in climate, culture and working conditions between Poland and the Caribbean was abrupt. The Polish astronomer was hired. His main responsibility was to keep the telescope in perfect order for the users who came from Princeton, Caltech and Cornell for their different observing programmes. This work enabled him to get to know many astronomers and also meant that during periods of low usage, the instrument was available for his own pulsar research. It was using this telescope that Don Backer, from the University of California, detected the first 'millisecond pulsar' in 1982, just a few weeks after the arrival of the Polish researcher.

COSMIC VAMPIRES

At the time, the discovery of a millisecond pulsar grabbed people's attention. Just think of it, a rotation period of 1.557 milliseconds is a good twenty times faster than the youngest and fastest of previously known pulsars, that of the Crab Nebula, which is 1000 years old. Yet PSR 1257+12 is apparently an old pulsar, 300 million years at least, or that's what the weakness of its magnetic field seemed to imply.

Did the whole theory of these strange objects need to be revised? No, judged two Americans from Princeton, Roger Blandford and Larry Smarr. In 1976, they pointed out the possibility of a neutron star being rejuvenated, becoming a pulsar once again, provided that it turned into a cosmic vampire. Since to do so it would need a victim and yet was not able to wander around space looking for one, it would need to have a victim that was close enough to just reach out and

The vampire pulsar

One of the theories on the origin of millisecond pulsars proposes starting with a binary system composed of an ordinary star (left) and a neutron star (right). Each star sits at the centre of the gravitational lobe created by its mass. Any particle of matter trapped in one of these lobes falls into the corresponding star.

As it ages, a star expands. It grows to the point where its outer layers enter the gravitational influence zone of the neutron star. This matter transfers its angular momentum to the latter, allowing it to start spinning very rapidly.

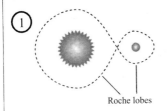

Roche lobes

Some of the matter from the vampire's victim forms a disc around the millisecond pulsar.

Planets could be formed from the disc fed by stellar matter. This is what must have happened to the millisecond pulsar PSR 1257+12, which hosts at least three small planets.

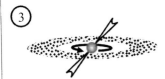

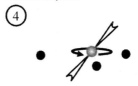

grab. It would have to be in a binary system of which the second component is a normal star. For a long time, the latter would evolve without bothering its bedfellow. Like any star of its type, it would go through the different stages of fusion and see its outer layers puff up and puff up. If it were near enough to its neighbour, there would be a moment when a part of the expanding matter enters the zone of gravitational influence of the neutron star, which would unabashedly suck in this free giveaway gas. In so doing, the gas would transmit some of its angular momentum to the cosmic vampire, which would restart spinning like a top, becoming a pulsar once more, and even surpassing the speeds of its adolescence.

After this first theoretical breakthrough, other variations for explaining the origin of millisecond pulsars appeared. There's one in which two neutron stars orbiting around one another attract one another and eventually collide – a case which could well apply to the binary system PSR 1913+16. It's also possible that the very energetic pulsar beams hit the stellar companion and make it gently melt until the matter pulled off falls on the neutron star and wraps around it. This could be what happened in the case of PSR 1957+20 in the Sagittarius constellation. Another scenario is a binary system in which a white dwarf, a dense and massive star, sucks in the expanding gas of its companion, a star in the giant phase. This flow of matter fattens up the white dwarf until it collapses under its own mass and transforms into a millisecond pulsar. Finally, another variation is of two white dwarfs that orbit around one another until they collide and give birth to a spinning neutron star.

In February 1990, a maintenance team discovered cracks in the structure that supported the platform hanging from the Arecibo telescope. The faulty part had to be fixed. For three weeks, the telescope was totally paralysed. It could do nothing but look at the sky directly overhead. For most programmes, the paralysis meant total inaction. But not for Alexander Wolszczan, who leapt at the chance and submitted a research proposal which could be satisfied with a handicapped Arecibo. His project was accepted, and the Pole was rewarded with many days and nights of reserved observing time, an allocation which is ordinarily unthinkable.

PSR 1257+12 IS THE LUCKY NUMBER

While most millisecond pulsars have been discovered in the plane of the Galaxy, Wolszczan had a fancy to look elsewhere. Day after day, night after night, he scrutinised the sky with his convalescent telescope. With 2000 samples per second over 32 different frequencies, he gathered no less than 64 000 samples per second. After many dozens of hours of observing, he was submerged by magnetic tapes. Luckily, the Cornell computing centre, with its extremely powerful computers, worked on decoding the data.

Alexander Wolszczan got the first decoded results in May 1990. At a glance, he discovered two pulsars, a binary, PSR B1534+12, which is in the constellation of Serpens, and a millisecond pulsar, PSR 1257+12, in Virgo, 1600 light-years from the Earth. At first, Wolszczan

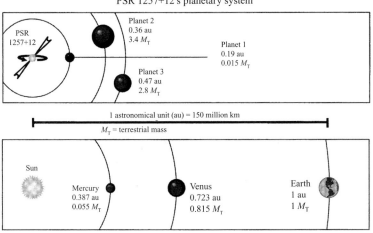

PSR 1257+12's planetary system

1 astronomical unit (au) = 150 million km

M_T = terrestrial mass

(The sizes of the stars are not to scale)

was particularly interested in the former, but he soon noticed that the latter seemed to conceal an even more fascinating secret. After having timed the radio emission of PSR 1257+12, he discovered a strange anomaly. None of the models that he constructed to explain the behaviour worked. This pulsar defied every expectation, until Wolszczan decided to consider the planetary hypothesis. It was autumn 1990. Because he wanted to avoid any upsets, he decided to reveal nothing for six months in order to be sure that the perturbation he observed was not the result of the Earth's movement around the Sun.

Six months later, the verdict came: the anomaly was not that of a badly corrected movement by the Earth. Wolszczan then appealed to his colleague and friend Dale Frail, who worked on the Very Large Array (VLA), a radio telescope composed of 27 dishes installed on a high plateau in New Mexico. He asked him if he could kindly carry out an independent measurement of the position of his pulsar. Frail did this. The signal was weak and the measurement difficult. The American got there anyway and confirmed his colleague's conclusions.

The two men published their discovery on 9 January 1992 in the review *Nature*. And six days later, Alexander climbed the podium during the Atlanta conference to present his work, at the very moment that the audience was reeling from the shock of Andrew Lyne's announcement. He had to overcome the pessimistic atmosphere, to convince others that his results were reliable. Fortunately for the Arecibo radio astronomer, the revolutionary periods of the two planets are 67 and 98 days respectively, that is, nothing at all like a period related to the Earth, while the masses of the two companions are 3.4 and 2.8 terrestrial masses. We were really dealing with featherweights.

There was another advantage that Wolszczan had over Andrew Lyne: it's easier to explain the formation of planets around a millisecond pulsar than around an ordinary pulsar. If the former is created as theoretically predicted, by sucking the gaseous blood of a star, there's a fair chance that the matter sucked in doesn't just nourish the vampire, but that some of it forms an accretion disc. There would then be a configuration quite similar to that which exists around stars as they are born, just before the formation of protoplanets.

The general feeling of the scientific community, though still stunned by Lyne's misfortune, was that Wolszczan's work seemed free of any instrumental or computing error. Moreover, two months later, Don Backer, the discoverer of the first millisecond pulsar, confirmed Wolszczan's data. All that remained was to gather other clues to confirm that the timing perturbations of PSR 1257+12 were really due to planets and not due to special starquakes at the surface of the neutron star.

Frederic Rasio, a physicist at the Massachusetts Institute of Technology (MIT), came up with an idea on how to do this. In 1992, he published an article in which he proposed the theoretical framework of an experiment. He noted that there was ratio of 3 to 2 in the orbits of the two planets: one carries out three revolutions while the other carries out exactly two. Therefore the two planets are regularly in conjunction. At each of these meetings, the two planets must mutually influence one another to the extent that their respective orbits are slightly modified. So, according to Rasio, this change ought to influence the pulsar and the configuration of its flashes. Wolszczan, who enthusiastically read the article, tried the test. He published his results, which confirmed Rasio's predictions, in the magazine *Science* on 1 March 1994. He also announced the likely existence of a third planet with a 25.3 day period and a mass about that of our Moon.

After the discovery of the treasures of PSR 1257+12, there have been other candidates for pulsars with planets. The most serious of the claimants is certainly PSR B1620–26. Andrew Lyne and his team first discovered it during a campaign in 1988, but over the years, others, like Steve Thorsett from Princeton, have joined them to better delve into the mysteries of this object. PSR B1620–26, located in the globular cluster M4, is a millisecond pulsar. So it can lay claim to an extreme stability and a precision just as impressive. It was quickly noticed that it's accompanied by a massive companion, probably a white dwarf, and then it was suspected that there is also another object, no more massive than 10 Jupiters and with an orbital period of a century.

The other pulsar planet candidate is PSR 1828–11. It too was detected by the British at Jodrell Bank in 1988, but its irregularities were only noticed three years later. According to the discoverers, there should be at least three planets in orbit around it, of 3, 8 and 12 ter-restrial masses. However, despite eight years of regular observations, Andrew Lyne remained cautious. There are errors that one doesn't want to repeat. He was wary of this young pulsar that was only 100 000 years old. Who knows if the neutron star hidden there is not just ag-itated by various somersaults due to hiccupping attacks capable of imitating the signals of several planets? But what bothered Andrew Lyne most of all is that there's still no strong hypothesis to expain the formation of planets around a normal pulsar. This caveat also applies to another candidate, PSR 0359+54, which had already hit the head-lines in 1979, and whose candidature has been defended since 1995 by the Russian radio astronomer Tatiana Shabanova, even though, in the eyes of most experts, the perturbations of PSR 0359+54 are due to background noise.

Despite its success, the quest for pulsar planets remains poorly known by the public, undoubtedly because these planetary systems are so unlike our own. Neutron stars have little in common with homely stars like the Sun. It seems certain that the pulsar beams would have made it impossible for life to develop on pulsar planets. It is nevertheless the case that Alexander Wolszczan's discovery constitutes a major turning point for astronomical science. By showing that planets can form in extreme conditions, it breathed new life into the search for other worlds and created the hope that there exist many more than were imagined earlier.

6   Brown dwarfs in the headlines

We saw in the previous chapters that cosmic fauna is incredibly diverse. But you don't need to be on first name terms with objects as exotic as black holes or pulsars to see this. Even the 'normal' star family has too many children to easily keep track of. Some way had to be found to classify this stellar family according to some sensible scheme. Every star is today identified by its colour (or spectral type) and by its luminosity (or absolute magnitude). The different spectral types have each been given a name, in fact a letter of the alphabet, and the sequence is now: OBAFGKM. The O stars are those whose surfaces are much hotter than any of the others. Some of them are well beyond 30 000 °C. Our Sun, at 5700 °C, is in the G class. The M class consists of the coldest stars, with mean surface temperatures of 2600 °C. Whatever their peculiarities and differences, all stars have, however, something in common: the thermonuclear fusion reactions of hydrogen that take place in their cores and which make them members of the main sequence, the club of normal stars.

There are so many stars undergoing nuclear combustion that it seems almost as ordinary as walking the dog. But those on Earth who try to control fusion, which is more powerful and less polluting than the fission used in nuclear power plants, know that it's a very difficult process to tame. Earning your wings for a stellar candidate requires several conditions to be fulfilled: a minimal mass, and at the centre, a temperature of at least 8 million degrees Celsius and a pressure of a billion atmospheres: anything less and the hydrogen nuclei refuse to fuse together.

It's enough just to utter the words 'minimal mass' for scientists to grab on to the term and try to nail it down, theoretically, of course, but also experimentally. For a long time, it was thought that M stars, also called red dwarfs, were the least massive and the coldest stars. The poor things! Not only does their name 'M' stand for a French five-letter word not to be spoken in polite company, but M stars are also puny by mass, which earns them their place at the tail end of the main sequence. You could say that they barely got their membership cards.

A FAILED STAR

In the 1960s, Shiv Kumar, an Indian researcher who settled in the USA, wondered about the theoretical minimal mass of a star. What was the limit below which nuclear fusion of hydrogen would not be ignited? His calculations showed a limit of 0.08 solar masses, which equals about 80 Jupiters (today's models indicate between 0.072 and 0.075 solar masses for similar proportions of elements to those in the Sun). Clearly, there was a big gap between the biggest known planet, Jupiter, and the theoretically smallest star. Was this gap filled with something, and if so, what could these objects of which we had no trace and which only existed on paper look like?

The heroine of Shiv Kumar's plot – which was to be taken up and developed by the famed Jill Tarter in the mid-1970s – was a failed star. A failed star at first forms like an ordinary star, consisting of roughly 73% hydrogen, 25% helium and about 2% of heavier elements. If its mass is high enough, close to 0.08 solar masses, then it manages to start deuterium fusion. It's even possible that it consumes some hydrogen, but only briefly – for not more than a few million years. After that, it snuffs out, cools down, and gently slides towards its destiny as a 'degenerate' object.

Without the temperature needed to dilate the gas and oppose it, gravitation then has free rein to compress the core of the star at will. But as the space available is reduced, the electrons go haywire. Quantum physics forbids them to occupy the same states as their neighbours. So they become agitated and speed up until they create enough pressure – quantum pressure, this time – to counterbalance the gravitational collapse. It's this strange equilibrium that earns the brown dwarf, like the white dwarf, by the way, the adjective 'degenerate'.

Shiv Kumar's hypothesis convinced researchers to begin the hunt for brown dwarfs. There's a lot that can be learnt from these degenerate and cold objects concerning both their overall physics, and also the complex meteorological phenomena (cloud and aerosol formation) that must occur on their surfaces, rather like the weather on the gaseous giant planets which they resemble. In short, learning more about brown dwarfs and especially about the least massive of them would be learning more about planets. And that's not all. Brown dwarfs can also teach us a lot about the process of star formation and complete the table of mass distribution, in other words, the numbers of stars per mass and per volume. Finally, these same brown dwarfs might hold the key to the mystery of the missing mass.

What is this missing mass? This concept is due to the Dutch astronomer Jan Oort. At the beginning of the 1930s, he was already known for having shown that the Galaxy turns around a centre 28 000 light-years away from our Sun, which places the Sun towards the edge of the huge Galactic disc. Very familiar with questions related to movements of the Galaxy, Oort was very interested in stars that succeeded in leaving its plane. These runaways need a considerable impulse to be able to distance themselves from the huge Galactic mass and its gravitational field. The higher their initial mass, the better their chances of going far. So, Jan Oort told himself that by measuring the escape speeds of these runaway stars he could deduce the attractive force of the Galaxy and hence its mass.

His results were surprising. The runaway stars behave as if the Galaxy is much more massive than it looks. Its visible matter, composed of stars and interstellar clouds, is insufficient to completely explain the gravitational influence that the Milky Way exerts on these runaway stars. So is there some invisible matter? And what could this be made of? Black holes, planets, unknown objects, tiny particles? It's a complete mystery. So, when brown dwarfs arrived on the scene of astronomical theory, it was thought that they could play a role in the problem of missing matter.

ASTROMETRY INVESTIGATES

Find brown dwarfs, that was the new catchphrase. But where should we look? When you're looking for a substellar mass, a cold and undoubtedly barely visible object, it's better to be methodical. Luckily, the hypotheses were sharpened up, the theoretical profile of brown dwarfs was refined and instrumentation improved.

It was very natural that astrometry (see Chapter 4) was attracted to brown dwarfs. As these are cold objects they are very faint, so an indirect method of looking for them could be invaluable. If brown dwarfs orbit around nearby stars, then astrometrical techniques should make it possible to see the oscillation that the former induce in the paths of the latter. It was already known that stars are often born in pairs. So what could have been more obvious than to look for brown dwarfs, which, even if failures, are stars around other stars, which themselves are normal? Moreover, depending on whether many or few were detected, this would give astrophysicists precious information for refining their models of star formation.

In 1978, Sarah Lee Lippincott, a collaborator of Peter Van de Kamp, wrote an article with Elliot Bergman on a star commonly called Gliese 623. From 455 images of it accumulated over 40 years, the authors concluded that it has a companion of 0.06–0.08 solar masses. It could have been a brown dwarf, or else a very low mass and very faint red dwarf. The latter hypothesis seemed the more realistic. In fact, the measurement made in the year 2000 by Damien Sagransan, of Grenoble, using the Élodie spectrograph installed at the Observatoire de Haute-Provence, gave a mass of 114 Jupiters for Gliese 623, while the maximum mass of a brown dwarf is 80 Jupiters.

The episode which followed that of Gliese 623 was undoubtedly one of the best known. It started in 1983 with the publication of an article by Robert Harrington, of the United States Naval Observatory, who was interested in a handful of particularly faint stars, including

VB8 and VB10. The 'VB' is in honour of the person who had studied these objects extensively in the 1960s: the American–Belgian George-Achille Van Biesbroek. These objects, barely as far as 20 light-years from the Sun, are intrinsically very faint, a property caused by their low mass. So they are perfect targets for astrometry. Not only do their low masses make them especially sensitive to the gravitational influence of possible companions, but also their proximity makes such oscillations more easily detectable.

Robert Harrington suggested that objects of a few hundredths of a solar mass accompanied VB8 and VB10. Being prudent, he avoided talking of brown dwarfs, feeling it was better to await a second opinion. Which is what Donald McCarthy and his colleague Frank Low, of the University of Arizona, associated with Ronald Probst, of Kitt Peak National Observatory, gave him in 1985. Their conclusions were mixed. Yes, they had noticed something around VB8. But no, they hadn't noticed anything significant around VB10.

In contrast to Harrington, who used astrometry, the three men used infrared detection, a wavelength domain where invisible objects can be revealed thanks to the heat they emit. Theoretically, a brown dwarf can attain 2500 °C at its surface while a planet like Jupiter hits a 'ceiling' of −173 °C. So it's very possible that it could give out a signal in the infrared, that it could be seen live. And this is all the more true if it's the companion to a star as close and as little luminous as VB8 and VB10.

The Americans used speckle interferometry, which was invented in 1970 by the Frenchman Antoine Labeyrie. Normally, to find a faint object, you lengthen the photographic exposure times, but this strategy has a major disadvantage: it lets atmospheric turbulence, shifts of air pockets, changes in humidity and temperature, pollute the photo by making the star light fluctuate. Speckle interferometry instead takes very short exposures, and uses a statistical approach to analyse the fluctuations of a star in order to reconstruct the image of the star that would have been obtained with a long exposure, without the problems from the atmosphere.

This is how, after having recorded thousands of photos at two infrared wavelengths, 1.6 and 2.2 microns, Donald McCarthy's team discovered a small luminous spot beside VB8. The properties of the object were determined: its luminosity was estimated at not more than 0.000 03 times that of the Sun and its surface temperature at about 1200 °C. As for its mass, a crucial parameter, the Americans estimated this to be between 0.02 and 0.05 solar masses, i.e. well below the hydrogen fusion limit. So there was a pretty good chance that it was a brown dwarf, or, with a bit of luck, even an exoplanet. The Americans unhesitatingly wrote in their article: 'These observations may constitute the first direct detection of an extrasolar planet.'

The joy was short-lived. In October 1985, during the first conference devoted to brown dwarfs, organised near Washington DC, three Americans, Michael Skrutskie, William Forrest and Mark Shure, unveiled the results of their research. They had made infrared observations of sixty stars located at less than 12 parsecs (about 40 light-years) from the Sun, and eight other stars in the Pleiades, with the sole goal of finding substellar mass companions. But they found nothing. One of two explanations was possible: either brown dwarfs are extremely rare, or else they cool down so quickly that they become too faint to be detected even in the infrared. For those who had thought that the discovery of VB8 B would spark off a miraculous harvest of low-mass objects, the Americans' result threw a dampener on things. And that was not all.

CANDIDATES APLENTY BUT NO WINNERS

In the autumn of 1986, a new article devoted to VB8 made waves. It was coauthored by Christian Perrier and Jean-Marie Mariotti, both astrophysicists at the Observatoire de Lyon. To make their observations, the two Frenchmen had the luxury of one of the purest skies in the world, that of La Silla, on a desert mountain in Chile. This is the spot chosen by the European Southern Observatory (ESO) to install several instruments, including a 3.6-metre Cassegrain telescope that had a new, particularly powerful, infrared, speckle camera.

Perrier and Mariotti chose two different wavelengths for their study: 2.2 microns – like McCarthy – and 3.6 microns, which was to enable them to estimate more precisely the effective temperature of VB8's companion.

The observing conditions were excellent. However, despite the mild weather, the Frenchmen didn't manage to see VB8's companion. There was no sign at all of a brown dwarf. After having gone though all possible sources of error, the two researchers concluded that there isn't a VB8 B further than 2.5 astronomical units from the main star, whereas McCarthy had seen it at 5 astronomical units.

VB8 B started to fade as a candidate, a process that accelarated when Skrutskie, Forrest and Shure revealed that VB8 was part of their stellar sample and that they had looked at it twice. They too regretfully announced that they had seen nothing special around this star, nothing that resembled a point-like object. In fact, their study had failed to show evidence of any companion in the range 0.04–0.08 solar masses around the 60 stars that they had selected. So, if brown dwarfs did exist, they must be much less luminous than had been thought, and the detectors of the time were simply not powerful enough to reveal them. The detectors had to be improved. A question remained: what had McCarthy seen? According to Christian Perrier, who analysed the data, it was very probably atmospheric perturbations that had confused the Americans' data. To remove any remaining doubts, Robert Harrington, who had continued his astrometric measurements of the star VB8, announced in turn that he too no longer detected anything. The brown dwarf candidate VB8 B had now definitively sunk into the cosmic night.

The race continued. The game was worth it. Whether or not brown dwarfs were eventually found, the sum total of knowledge acquired would by far justify the time spent chasing these stellar ghosts.

While some were waiting for more powerful detection instruments that would be able reveal these cold stars, other researchers came up with incredibly cunning methods of making do with what was already around. So, during the northern summer of 1987,

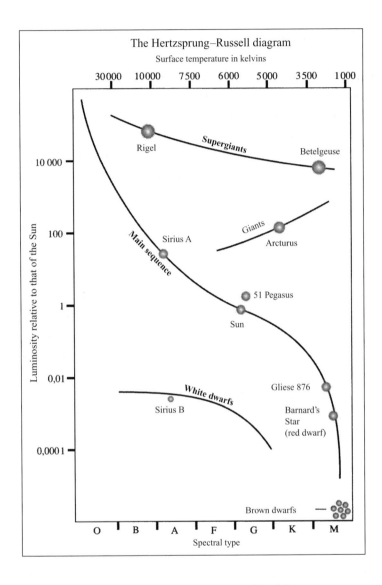

The Hertzsprung–Russell diagram

Surface temperature in kelvins

Benjamin Zuckerman, of the Univesity of California in Los Angeles, and Eric Becklin, of the University of Hawaii, published the results of some astonishing research. Inspired by an original idea from Ronald Probst, of Kitt Peak National Observatory, they focused on fourteen known white dwarfs in the hope of finding brown dwarfs orbiting them. This was a pretty weird idea since white dwarfs are degenerate objects (like brown dwarfs), hot and hyperdense balls, as massive as the Sun but concentrated in a sphere the size of the Earth. A white dwarf is made when a star like the Sun can no longer feed its nuclear reactions. It releases its outer layers, revealing a very massive core in a volume equal to that of the Earth.

But why should you search for brown dwarfs around white dwarfs? Could it be because both are dwarfs? Yes, there's something in this since if a brown dwarf has survived the chaotic end of its main star (the further away it is, the better its survival chances are), it would be easier to detect. In fact, while their high temperature causes white dwarfs to emit radiation in all sorts of wavelengths, though mainly in the visible and the ultraviolet, in contrast, they don't emit much in the infrared. Yet infrared emission is a prominent property of brown dwarfs. So it's enough to focus a good detector at a well-chosen wavelength on a white dwarf. If it emits an excess of infrared, you can guess at the presence of a brown dwarf in orbit around it.

Zuckerman and Becklin started their hunt in mid-November 1986 and continued it right through to January 1987. They had at their disposal the InfraRed Telescope Facility (IRTF), a 3-metre diameter infrared telescope, sheltered by the domes of the Mauna Kea Observatory, at Mauna Kea in Hawaii. Nights passed and they detected nothing above a mass of 0.015 solar masses, the sensitivity threshold of their detector. Far from being discouraged, the two men announced a new campaign, aimed at 'younger' dwarfs. Why not? Maybe the new campaign would be just what was needed to find the coveted Holy Grail!

Zuckerman and Becklin didn't give their colleagues much time to cogitate upon this. Three months later, they came back into the limelight with a new article in the review *Nature*, in which they revealed that they had found a white dwarf called Giclas 29–38, which was located in the constellation of Pisces and which seemed to emit an infrared excess. If the source of the emission was, as the authors thought, a spherical body with a surface temperature of about 900 °C, then its radius had to be about a sixth of that of the Sun. They determined that the supposed brown dwarf had to be located about 5 astronomical units from its main star, and that its mass had to be 0.04–0.08 solar masses.

Prudently, the American team carefully avoided declaring the nature of the object. Everything had to be confirmed. And as self-help is always the best, Zuckerman and Becklin, joined by Alan Tokunaga, of the University of Hawaii, continued their investigations. In September 1988, they delivered new conclusions threaded with prudence and caveats. According to them, their new measurements made the existence of a spherical object uncertain. If the distance between Giclas 29–38 and its companion was above 5.6 astronomical units, then it was possible that the latter had shifted along its orbit, that it had passed behind the star. This game of hide-and-seek would explain the change in the infrared emission. Or instead of a brown dwarf, there could be a disc of dust heated up by the main star. The American James Graham and his Caltech team pored over the latter hypothesis and in 1990 published a series of arguments which strongly supported it.

In contrast, another candidate found during the same search seemed to better resist the critics. Also detected by an infrared excess, it orbits a white dwarf named GD165, at a distance of about 120 astronomical units. The good news was that its temperature was not above 1800 °C, and several clues existed that excluded the possibility of a background galaxy. As for the nature of its companion, everything pointed to it being not a disc, but a compact object. Only its mass raised doubts: it was estimated at about 0.08 solar masses. Is it a brown dwarf or a small star seeking a quiet life? This is still an open question today.

Three Americans, William Forrest, Michael Skrutskie and Marke Shure, the same three who had recently contributed to the rejection of VB8 B, after an infrared search around about fifty-five stars located in the solar neighbourhood, declared that they had found a companion to the star Gliese 569, which was colder than the coldest stars known. Despite this, the object had a slightly higher luminosity than that of standard low mass stars. Either this was a young star rapidly evolving towards the main sequence – the grown-up world for stars – or else it was a brown dwarf at the beginning of its cooling phase. A study published in 1990 by Todd Henry and Davy Kirkpatrick concluded that the mass of Gliese 569 B was about 0.09 solar masses, suggesting a preference for the faint star hypothesis.

At the end of 1988, the American Wulff Heintz caused a storm by introducing Wolf 424, a binary system composed of two dwarfs, the orbital period of which is 16.2 years. Using the astrometrical method, he followed the stellar couple for many years. In 1972, he had already underlined the fact that the members of this peculiar binary should have had masses close to the hydrogen ignition limit, on the borderline between red dwarfs and brown dwarfs. Sixteen years later, supported by 715 images taken over 50 years and covering three orbital periods, the Sproul Observatory astronomer judged that he was able to say confidently that the masses of the companions were 0.059 and 0.051 solar masses. Again, they were in fact red dwarfs.

Brown dwarfs were playing decidedly hard to get. It was as if they didn't exist. Or perhaps they happen to be rather unsociable and they don't care much, for star formation reasons, for the company of normal stars. After all, since they're born from the fragmentation and the collapse of an interstellar cloud, they don't need anybody's help to be born, in contrast to planets which owe their existence to the remnants of gas and dust which surround a protostar. So the next attempt to find them centred on young star clusters, with the idea that since a young brown dwarf is more luminous, it should be much easier to pick out. In contrast to previous work, which looked for the companions of main stars, the new searches sought independent brown dwarfs located in the small stellar bunches called clusters. While these sorts of searches are less precisely targeted, their yields can be big. By looking more widely, there's a small chance of seeing more. Both the general approach and the method itself are different. As long as you're looking for the companions of normal stars, it's possible, or so it was thought, to alternate the indirect astrometrical method and infrared observations. However, when searching in star clusters only the infrared method is usable; it is the only domain in the electromagnetic spectrum where brown dwarfs might allow themselves to be observed.

Several teams started off on this path. For example, the Pleiades, a cluster located about 400 light-years from the Sun and about 70 million years old, were looked at. Late in 1989, Richard Jameson and Ian Skillen of the University of Leicester in England asserted that they had found five objects in the Pleiades the masses of which ranged between 0.06 and 0.08 solar masses. But it wasn't possible to be certain what these objects were. On one hand, the masses were close to the limit of the lightest normal stars, the ubiquitous red dwarfs. On the other hand, there was no guarantee that these objects really belonged to the Pleiades cluster, that they weren't just background stars further away. The only way to be sure would be to carry out the astrometrical measurement of their proper motions in order to see if they corresponded to those of the cluster. This required years of effort. And even if you could show that a suspect object was really in the cluster, how could you be sure that it was really a brown dwarf? There is only one solution: check out the theoretical models which establish links between the luminosity of an object, its age and its mass. However, due the lack of data, the models are still quite approximate and are not good enough to produce clearcut decisions.

A bit more patience was needed. For five years, announcements of discoveries followed one after another, each of which resulted in fuzzy conclusions. Again and again there were the same doubts about the nature and the real distance of the objects. But instruments got better, became more and more powerful and able to detect fainter and fainter infrared sources.

FINALLY SOME CONFIRMATIONS

Pittsburgh, June 1995. Hundreds of astronomers gathered at the annual meeting of the American Astronomical Society. Those who listened to the talk given by Gibor Basri of the University of California at Berkeley learned of the existence of a new brown dwarf candidate,

Palomar Pleiades 15 or, more informally, PPl 15, a discovery, as its name indicates, in the Pleiades cluster.

To be honest, this was really a confirmation. The first of its kind. The real discovery of PPl 15 was by John Stauffer of the Harvard Smithsonian Center for Astrophysics who, with his team, had worked through this cluster since 1989 looking for independent brown dwarfs. Thanks to the infrared and to a CCD detector, a sort of electronic eye able to transform a flow of photons into a sequence of zeros and ones, a binary language understandable to computers, he had brought to light six interesting candidates, some of which had particularly faint luminosities. Among them was PPl 15, the mass of which seemed to be about 0.06 solar masses. But there remained a doubt about whether it really belonged to the Pleiades cluster. To be sure, its shifts in the sky, its proper motion, had to be measured, to see if it agreed with that of its bright neighbours. However, measuring proper motion is all the more tricky when it comes to measuring such a faint and difficult-to-follow object. So, the article's authors advised great caution to their readers.

Rather than bothering about these distance problems, Gibor Basri chose in preference another method, just recently made available in the great astronomy marketplace, a technique first described by Rafael Rebolo, a researcher from the Institute of Astrophysics of the Canary Islands. Rebolo claimed that it was possible to apply a test to distinguish a brown dwarf from a red dwarf. In any star that forms from an interstellar cloud, there's a lot of hydrogen, some helium, and also a few other elements in small quantities. Among these is lithium, which normally disappears quickly, a victim of thermonuclear reactions. But brown dwarfs don't undergo hydrogen fusion. Obviously, they should keep their small initial dose of lithium.

All that was left to do was to search for the lithium and for this spectroscopy was required, a technique that makes it possible to decode the light that comes from the stars and to deduce the presence of one or other chemical element on its surface (we'll say a lot more about this in the next chapter since this technique is intimately linked with the discovery of the first exoplanets around ordinary stars). So it's possible to decode the light that arrives on the Earth and to learn more about the nature of the star that sends us the light. This is exactly what Gibor Basri did: he applied the lithium test to PPl 15.

I remember meeting Gibor at the University of Hawaii. During an informal discussion, he had revealed to us that he was studying PPl 15 and that he had successfully applied the lithium test to it. This was clearly a major breakthrough. But a few days later, Basri asked us to forget what he had said. He had carried out a second test to be sure, and the spectrograph hadn't shown any lithium at all. It goes without saying that we were all very disappointed. We had to again resign ourselves to being patient. And just as we were getting reaccustomed to the idea, Basri withdrew his latest remarks and told us that an error had slipped into the second test. Lithium was well and truly present. Whoever claimed that the path of science is straight and narrow?

While PPl 15 was hitting the scientific headlines, another brown dwarf joined it in the limelight after an article was published in *Nature* on 14 September 1995. Called Teide 1, in honour of the volcano on which the Observatory of the Canaries is built, it too is situated in the Pleiades cluster. Its discoverers were Maria Zapatero Osorio, Eduardo Martin and Rafael Rebolo, the inventor of the lithium test, a test that he had, by the way, performed on two well-known candidates, Gliese 569 and Wolf 424, without detecting lithium in either of these.

Teide 1 achieved the feat of being still less luminous than PPl 15. Its mass was originally estimated at 0.07 solar masses. This was extremely close to the lower limit for red dwarfs. Luckily, it brilliantly passed the lithium test and its mass was measured to be less than 0.05 solar masses. It was undoubtedly a brown dwarf.

Despite everything, PPl 15 and Teide 1 never had the same success in the media as Gliese 229 B. Gliese 229 B is the true heroine of the year 1995, it was she who definitively confirmed the existence of brown dwarfs. Located at some 19 light-years from the Sun, she's not single, but married. She's accompanied, for better or worse, by the star Gliese 229. Her existence, revealed by Tadashi Nakajima and his colleagues from Caltech and Johns Hopkins University, was announced during the famous Florence conference in 1995 when 51 Peg's companion made its official entry into the astronomical world. Which goes to show that those were heady days for the search for substellar companions.

Gliese 229 was part of a sample of a hundred stars chosen according to several criteria: their closeness to the Sun, their low mass and their old age. The first two criteria make it easier to detect a companion. The third makes it possible to avoid any error in the identity of the object. If a brown dwarf accompanies an aged star, it will have had time to cool down and should be less easily confused with a faint star like a red dwarf. The researchers used the telescope of Mount Palomar, which had been equipped with an infrared camera and also with a coronagraph making it possible to simulate an eclipse of the star and to observe its neighbourhood without being blinded.

The Americans looked at Gliese 229, a very faint red dwarf, on 27 and 29 October 1994. Once the coronagraph was in action, they saw a sort of outgrowth at the edge of the star. It really seemed like the main star was hiding something in its side. A brown dwarf? Possibly, but it was too early to tell. Some things just can't be rushed. If one year later the apparent distance between the two objects hadn't changed, then it would be possible to conclude that it was a true couple.

The verification photo was planned for 17 November 1995. The camera used was the most prestigious in existence: the Hubble Space Telescope itself. It confirmed that Gliese 229 and Gliese 229 B formed a very beautiful couple. So, was it a brown dwarf? Yes, without the shadow of a doubt. The spectral measurements confirmed it: they showed the signature of methane, a molecule that can't survive at temperatures above 1250 °C. Yet no normal star is supposed to go below 1450 °C.

So Gliese 229 B was called the first cold brown dwarf, with an official validity certificate. At 44 astronomical units from its star, i.e. the equivalent of the distance between the Sun and Pluto, it has a surface temperature of about 700 °C, and its mass is between 0.02 and 0.05

solar masses. Success at last! The press was feverish with delight! In February 1996, the review *Ciel & Espace* (Sky & Space) when alluding to palaeontological searches, carried the headline: *Étoile-planète: la découverte du chaînon manquant* ('Stars-planets: the missing link').

This time it was not another false start. Brown dwarfs were no longer just theoretical, but an experimental reality. The missing link well and truly exists. However, this doesn't stop it from being enveloped by many mysteries. Astronomers wanted to know more and to learn the secrets of the formation of brown dwarfs, their composition and how common they are. And the best way to do this was again to find yet more candidates, using all possible strategies: infrared detection in young stellar clusters like the Pleiades, infrared or astrometric searches for companions around ordinary faint stars, and also, what we haven't yet mentioned, the study in the infrared and the visible of large portions of the field of the Galaxy. This third method is by far the least targeted of all. It does not involve looking at the edges of a star or even a small group of stars. Instead, it's about looking widely and sweeping through vast tracts of sky without being biased by the positions of the objects that one wants to discover. It goes without saying that such a procedure requires a lot of equipment and time, and so necessarily money. This is why these searches often involve several programmes with diverse aims. While this method is cumbersome and expensive, it has a significant advantage: it gives precious information on the frequency of different objects in the Galaxy.

In the middle of the 1980s, scientists like Patricia Boeshaar, of the University of Drew, adopted this method without arriving at any conclusive results. The instruments of the time were simply not powerful enough for this sort of detection. Ten years later, instrumental limits were much less of a constraint.

It was in 1987 that Maria Teresa Ruiz of the University of Chile started her programme of studying the Galactic field. She wasn't looking for brown dwarfs but for white dwarfs. With a 1-metre telescope at La Silla, she photographed large portions of the sky, looking for hyperdense and faint objects. Her technique consisted of comparing two photographs of the same celestial region taken ten years apart and keeping only the very faint objects that had moved. Those that have moved in the sky plane must be close to us and so, if they look faint, they must be intrinsically faint.

During a comparison session, in 1997, the scientist noticed that one point had shifted in the Hydra constellation. Following procedure, she requested the spectrum of the object to find out what it was. To her great surprise, the spectral signature was not that of a white dwarf. To be honest, she had never seen this sort of a star. This is unsurprising as it was a solitary brown dwarf, as the lithium test soon confirmed. The object, located at about 33 light-years, was named Kelu-1. Its mass is about 0.075 solar masses.

Other programmes studying the Galactic field were begun. Among these is the European DENIS (an acronym of DEep Near-Infrared Survey) of which one the most active organisers is the Frenchman Xavier Delfosse. The DENIS programme is to carry out the first digital mapping of the astronomical sky in the Southern hemisphere, on a large scale and in the infrared. It's planned to catalogue 100 million stars – some of which may be brown dwarfs – and 250 000 galaxies. Its first success came in 1997 with the discovery of three brown dwarf candidates, all of which successfully passed the lithium test.

Suddenly, the Galactic field zealots found themselves with four brown dwarfs, Kelu-1 and the three DENIS ones. The latter are very different from everything that had been discovered earlier. With the exception of GD165 B, the companion of a white dwarf, all the other candidates flirted with the least-massive stars in the M class, the red dwarfs, showing spectral signatures typical of titanium or of vanadium. The latter elements don't exist in the DENIS brown dwarfs or Kelu-1. Undoubtedly this is a question of temperature.

In the USA, the 2MASS programme, run by Davy Kirkpatrick and Adam Burgasser and focussed on the Northern hemisphere, announced in 1998 the discovery of several cold objects resembling the DENIS ones and Kelu-1. It could no longer be regarded as a coincidence. There does exist a sort of brown dwarf family that is relatively homogeneous. This was all that was needed for the experts to propose a new stellar class, the L dwarfs, whose average surface temperature would be between 2000 °C, the lower limit for M dwarfs, and 1250 °C. The system of star classification now included OBAFGKML, with the L class containing the brown dwarfs and possibly also, though this remains to be confirmed, especially light normal stars.

Searches in the Galactic field continue. 2MASS and DENIS are continuing to work in the infrared, while SLOAN, a new German–American–Japanese project, looks in the visible wavelength domain. All of these searches are aimed at a common goal: to find brown dwarfs that resemble Gliese 229 B, the unusual spectrum of which has the famous methane absorption line, an element well known to planetologists since it's present in the atmosphere of certain of our planets like Jupiter. If it's possible to discover brown dwarfs of this sort, we would no doubt be able to add an essential piece to the theoretical scaffolding that we use to try to understand how planets and brown dwarfs differ and how they are alike.

In 1999, all the efforts were rewarded. Many brown dwarfs were unearthed and all showed a methane absorption line. This was the moment to create a new star class. One now talks of T dwarfs for these methane objects, the temperature of which is lower than 950 °C. In contrast to L dwarfs, which one could imagine might still include a few normal but low mass stars, just on the borderline of the hydrogen burning limit, the hitherto unheard of T category can only contain brown dwarfs, with masses in the range of 0.03–0.06 solar masses (or, if you prefer, 30–60 Jupiters).

CLOSER AND CLOSER TO PLANETS

Getting back to the searches in open stellar clusters, the team from the Institute of Astrophysics of the Canary Islands led by Maria Zapatero Osorio and Rafael Rebolo, and also by Jérôme Bouvier of the Observatoire de Grenoble, set sparks flying. Between 1995 and 1998, these researchers delved into the Pleiades and unearthed several brown dwarfs whose masses go down to 35 Jupiters. They then looked at another cluster, that which is centred around the star Sigma Orionis, at about 1100 light-years from the Sun. This is younger than the Pleiades. No star there seems to be older than 8 million years. If there are any brown dwarfs there, they are younger, not yet cooled down, so they are easily detectable, which should make it possible to detect objects that are lighter still. This is the crux of the matter: to use this visibility to descend as low as possible on the mass scale and to see if brown dwarfs are able to tread on planets' territory, to flirt with masses of 10 Jupiters.

The harvest was generous. Dozens of candidates fell into the Spanish nets, with sometimes unbelievable masses. On 4 October 2000, in the review *Science*, the Spaniards of the Institute of Astrophysics of the Canaries, allied with American and German researchers, announced the discovery of eighteen planetary mass objects (between 8 and 15 Jupiters), discovered floating freely in the stellar cluster Sigma Orionis without being attached to any central star. The team decided to use the word 'planets' to define these objects after having detected the spectrographic signatures of molecules that can't exist on the surface of a star in their atmospheres. So they are particularly cold objects, and this is all the more astonishing given that the Sigma Orionis cluster is known to be young, 5 million years old at the most.

Similarly, we should also mention the work of the Englishmen Patrick Roche and Philip Lucas, who, in March 2000, announced the discovery of dozens of astonishing objects in Orion's Trapezium. These celestial bodies couldn't be older than 2 million years, while their masses, in some cases, are below 13 Jupiters.

So, are these planets or mini-brown dwarfs? This is a question that has continued to trouble the experts in the last few years to such an extent that the International Astronomical Union nominated a special working group to deal with this question of semantics.

Before the discovery of brown dwarfs, 51 Peg b and its sisters, the dividing line between planet and star seemed clear. Our Solar System unambiguously gave us the definitions we needed. A star is a body that is made from an interstellar gas cloud that gathers together under the effect of its own gravity until it forms a ball and then, if it has enough mass to set off nuclear fusion reactions, a sun. In contrast, a planet, whether telluric like the Earth, Venus or Mercury or a gas giant like Jupiter, Saturn, Uranus or Neptune, forms by the accretion of bigger and bigger grains that are made from the protostellar disc, or in other words, from the matter that the star has not used for itself. Clearly, according to this definition, a planet can only be made from what is left over from the birth of a star.

But how does this way of seeing things fit with the latest discoveries? How should we classify the objects detected in Sigma Orionis which are of such low mass that they resemble our planets, yet float freely without being attached to any star? Personally, while accepting that it would be premature to say anything given the unfinished nature of our research, I tend to favour a definition that, rather than being linked to the mass of celestial objects, is based on their formation processes. My feeling for the objects discovered by my Spanish colleagues is that we have here a population that constitutes the lower limit of the star family. Several astrophysicists like Lynden Bell, Martin Rees and Alan Boss have shown that the star formation process, i.e. the fragmentation of an interstellar cloud, could yield some objects as light as just a few jovian masses.

Is this enough to be convinced that these are not planets? To be honest, no. Several clues – which we'll consider later – make it possible to imagine that several gas giants can form from a single protoplanetary disc, before one (or several) of them, in a great gravitational billiards game, is ejected from the system into the interstellar void. Is it possible that this is the case for the bodies detected by the Canaries team? In that case, they would truly be planets, in the classical sense of the term. But how can we be sure? No way of deciding exists for the moment. At best, we can be surprised to note that my Spanish colleagues' objects are rather massive for planets, at 5–10 Jupiters. But in the gravitational billiards game, it's especially the light objects, the small planets, that have the best chance of being ejected.

To sum up, uncertainty and puzzlement still dominate the semantics. However, it's possible that the objects discovered in Sigma Orionis and in Orion's Trapezium will one day cause a revolution in our criteria for the classification of celestial objects and in our theories of how substellar mass objects form.

Since at the moment brown dwarf candidates only come in handfuls, we're forced to accept that their diversity is surprising. Cold or hot, heavy or light, sociable or solitary, rich in something, poor in something else, they come in all varieties. However, there is one activity that they don't seem to look on kindly, which is to be a star's companion. It's as if they prefer to leave that role to planets. This is why the discovery in 1988 of the companion of HD 114762, our next episode, would cause so much ink to flow. With its mass of at least 11 Jupiters (and there's a 50 : 50 chance that it's above 13 Jupiters), we don't really know what to think of it. Brown dwarf or planet, it's hard to say. Doubts linger today and will continue until future more powerful instruments, like the interferometers, make more precise measurements possible. Nevertheless, the discovery of this strange object will have heralded a new era, marked by the discovery of multiple exoplanets thanks to the method of radial velocities.

7   Sirens in the Cosmos

On the morning of 5 August 1988, the readers of the daily, *Libération*, read on page 19 a headline which was intriguing, to say the least: 'They see planets everywhere'. This was intriguing and somewhat enigmatic. You had to read the introduction to better understand: 'Two astronomers, a Canadian and an American, claimed the day before yesterday to have discovered new solar systems. Shivers. And doubts.'

Baltimore is in the American state of Maryland and is home to the Space Telescope Institute, the control centre of the brand new Hubble Space Telescope. In August 1988, the town welcomed the twentieth General Assembly of the International Astronomical Union for several days. Nearly 2000 astronomers were expected from 54 different countries. This was the moment that two teams of researchers chose to announce their discoveries. On the one hand, there were the Canadians Bruce Campbell and Gordon Walker, from the University of Victoria (British Columbia), and on the other, a team led by David Latham, the American from the Harvard–Smithsonian Center for Astrophysics.

Since the beginning of the 1980s, Campbell and Walker had followed about twenty nearby stars, looking for substellar mass companions: brown dwarfs, of course, but also hopefully giant planets. Their quest seemed to have succeeded. Nine of their stars showed behaviour that could well have been due to such companions. According to the two researchers, it was very unlikely that the objects were brown dwarfs, because, they argued, if they were brown dwarfs, then they would have been detected by astrometrical techniques. Therefore, instead they had to be exoplanets with masses between 1 and 10 jovian masses. But Campbell and Walker were cautious and stressed that they had to obtain more data and make more measurements. Only the star Gamma Cephei had been observed long enough to allow experts to trace a complete orbit of a supposed companion.

The second team to hit the headlines during that conference only presented a single star, HD 114762, located 90 light-years from the Earth, in the constellation of Hercules. With a mass equal to that of the Sun, it is of a ripe old age, about 8 billion years. It would have never been thought to be very special if it hadn't been for a light perturbation agitating it. It was David Latham's team that had the major claim to the discovery. While we, meaning the team that I formed with Gilbert Burki, of the Observatoire de Genève, had the pleasure of being associated with it.

THE DISCOVERY OF HD 114762

A few weeks before the Baltimore meeting, I got a fax from David Latham in which he asked us for some details about HD 114762. Our American colleague thought that it was oscillating under the influence of a companion of substellar mass and wanted the opinion of a second group. This is understandable: the history of exoplanets is sufficiently full of rejected candidates that there's no need to add another to the list.

HD 114762 was known to us. Like David Latham, we were using it, as well as a hundred other stars, to calibrate our instrument called Coravel, a spectrograph installed on the 1-metre telescope of the Observatoire de Haute-Provence. We had to be sure each time that we used the spectrograph that it was perfectly focussed and calibrated. We had chosen the reference stars for their great stability. They were supposed to live alone, without stellar companions to perturb them. But that didn't prevent us from being careful. The machines that had concluded that those stars were stable were less powerful than ours, so it was quite possible that in front of our very eyes some of them would reveal that they were living with a partner. So we had to pay attention to their behaviour and see if they acted in unseemly ways, detectable by measuring their radial velocities.

In contrast to astrometry which measures stars' speeds (their proper motions) by following their shifts across the sky background, the radial velocity method measures stars' speeds along our line of sight. This technique can tell us whether a star is receding or approaching us and at what speed, thanks to the information contained in its light. And it's due to this same light that we can discover, if we have sensitive enough instruments, whether a star is animated by a light oscillation induced by the presence of a companion.

In 1985, in an article in which we presented the behavioural status of our reference star sample, we noted the radial variations of HD 114762 and mentioned the possibility that they were caused by a nearby object. However, we didn't we really try to clarify the point. As I said, these stars were just tools that we used in order to properly run our real research programmes, fifteen in all. For example, we had committed ourselves to complete the astrometrical data collected by the astrometric satellite Hipparcos on more than 40 000 stars by measuring their radial velocities. We were also looking for binary stars in different stellar agglomerations such as the globular cluster 47 Tucana and in open clusters like the Pleiades.

Also my collaborator, Antoine Duquennoy, and I had undertaken the task of measuring the radial velocities of close stars similar to our Sun in order to obtain a precise as possible census of the number of them that had partners. Our initial sample included 269 stars. In the end we only kept 164, those which gave us the most reliable data. Our conclusion was that most of them live with company. The article, published in 1991, remains one of the most cited from that year. I often say that it was slave labour, long and meticulous. But it seems that it was useful to many others as well as us, since it's thanks to that work that we ended up becoming curious about the question of low mass companions.

When David Latham asked us for details about the star HD 114762, we were studying our reference star sample to look for possible binary stars that could be interesting. The message we received from our colleague Latham at the beginning of 1988 encouraged us to once again delve into our data. We then had about 80 measurements of HD 114762. They were easily enough to confirm the presence of a substellar companion orbiting it at a distance similar to that between the Sun and Mercury with a mass of at least eleven times the mass of Jupiter. And if we say 'at least', it's because the radial velocity method can only give a statistical estimate of the masses of the companions that it detects. This is why there are doubts about the true nature of HD 114762 B, the estimated mass of which lies at the limits between two worlds, that of the brown dwarfs and that of the planets.

Despite this, during his presentation at the Baltimore conference, David Latham chose to openly speak of a planet. In contrast, the article that appeared in the review *Nature* instead mentioned a brown dwarf. Why the change? Simply because following the General Assembly, David Latham, his colleague Tsevi Mazeh and myself got together to write the article. The question of HD 114762 B's identity couldn't fail to be discussed. And while Latham argued in favour of the planetary hypothesis, Tsevi Mazeh and I expressed our clear preference for a brown dwarf hypothesis. Very democratically, Latham accepted the majority preference.

The discovery of HD 114762's companion was, without a doubt, a major event. Not only did it stimulate, at the time, the hopes of hunters of substellar mass objects, but it also confirmed the competitivity, in the field, of the radial velocity method, since, just like Latham's team and our own, the Canadians Campbell and Walker had used this technique to find their nine substellar mass candidates (which, in the end, succumbed to later observations).

THE ERA OF SPECTROGRAPHY

According to legend, one day in 1666, Isaac Newton was walking in the market place when he saw a piece of glass shaped as a prism. He bought it and returned home. Once at home, he shut himself up in a room, closed the shutters and made a small hole in one of them so that a beam of light could enter. The scholar then placed the prism along the path of the trail of light. The beam of light was suddenly transformed into a magnificent multicoloured band: red, yellow, green, blue, indigo and violet. By imitating the drops of rain which refract (disperse) the rays of the Sun, the piece of glass made a rainbow, or if you prefer, in expert's jargon, an artificial spectrum of solar light.

But to go from artificially creating a phenomenon to explaining it sometimes takes centuries of work. Newton lacked quite a few of the facts needed to describe in full what he observed in the dark room. In contrast, for twentieth century physicists, it was an almost trivial problem. Sunlight, like that of all stars in the Universe, is a composite light. It's the result of adding up an infinity of visible colours to which you have to also add the wavelengths that escape our eyes, for example the infrared and the ultraviolet.

So the solar spectrum is extraordinarily rich. Most of the time, we don't notice its subtle mixture. But it is sufficient to pass light through a refractive medium – a prism for example – to show this variety. When solar light enters a glass prism, each colour follows its own trajectory, according to an individual angle, and ends up participating in the creation of a rainbow.

But the Sun's spectrum coming out of a prism not only contains a continuum of visible, infrared and ultraviolet colours. If you look really closely, you might notice that it also contains hundreds and hundreds of lines which vary in darkness, thickness and sharpness. These are called absorption lines. They're the signatures of different chemical elements that are found on the surface of the Sun. And what is true for our star is also true for any other celestial object, star or planet, in the Universe. This is why absorption lines are a very valuable investigative tool for astrophysicists.

In order to understand the phenomenon, we have to make a quick detour through the heart of the matter: and the matter is matter itself. At the centre of an atom is a nucleus, composed of protons (with a positive charge) and neutrons (without a charge). Then, far from and around this central mix, we meet electrons organised in superimposed layers, placed in orbits at varying distances from the nucleus. Where the electrons are depends on their numbers. If the lowest orbit can't hold any more electrons, then the excess ones are distributed into a higher orbit, and so on. Crucially, the electrons are strictly forbidden from occupying a state that lies between two orbits. There are no subtleties. The orbits correspond to discrete (discontinuous) amounts of energy. This is one of the strange requirements of the world of quantum physics.

This doesn't stop electrons from passing from one orbit to another, higher one. But to make this small quantum leap, it has to receive some external energy, such as that which can be provided by a photon, in other words, by a particle of light. And it's also necessary that the photon has the right amount of energy. Otherwise the electron can't make its jump. But even then, that doesn't mean that it will stay in the higher orbit for long. Electrons are made such that they always like to occupy the lowest orbit possible, the lowest energy level possible. So, after a brief instant of excitement, the electron returns to its previous state by 'spitting out' a photon of the same energy as that which enabled it to make the quantum leap.

The absorption of photons by electrons is the basis of the dark lines that permeate the Sun's spectrum and the spectra of all stars. It was Gustav Robert Kirchhoff (1824–1887) who was the first to understand that these lines correspond to the signatures of chemical elements. Each has its own. That of iron is nothing like that of oxygen, which looks nothing like that of copper, of tantalum, of palladium, etc.

You have to identify each of these signatures before being able to detect them in stellar spectra. This is laboratory work, during which each known element is isolated, heated up and its characteristic spectral signature recorded. The end result is a catalogue that makes it possible to identify the lines of any complex spectrum.

In 1868, Anders Jonas Ångström (1814–1874), a Swedish scientist, produced a very high quality solar spectrum. You could distinguish no less than 1000 absorption lines, with an error margin of 1 angstrom (Å), i.e. just a ten-billionth of a metre, in the light's intrinsic wavelengths. In the same year, the Englishman William Huggins (1824–1910), who is often considered the father of modern astrophysics, made a primordial discovery. A self-taught man, with no university degrees, he set up an observatory in his garden near London and started experiments in spectroscopy in 1862. He described the spectral signature of different atoms like nitrogen and oxygen, and directed his telescope towards fifty or so bright stars in order to collect their spectra and to decode their absorption lines. In this way he concluded that the atmosphere of the star Aldebaran contains sodium, magnesium, calcium, iron, hydrogen, mercury, etc.

Thanks to this work, Huggins proved that chemistry is the same everywhere in the Universe. The elements are the same everywhere and the laws that govern them are just as universal. But the Englishman did yet more for science in general and for astronomy in particular. In studying the spectrum of Sirius, the brightest star in the sky, he noticed that one of the characteristic absorption lines of hydrogen was not at exactly the same place that it occupied in the laboratory. Sure, the difference was small – it was just 1.09 Å and not a bit over – but William Huggins was convinced that it was not an instrumental error. There was a real shift and this shift was the proof that Sirius is moving and that it's receding from the Sun.

His reasoning was perfectly correct. There is truly a relation between the positions of lines in the spectrum of a star and its movement – radial, that is – with respect to the Sun. Before arriving at this brilliant conclusion, Huggins had to absorb the theoretical development by two men, the Austrian Christian Doppler (1803–1853) and the Frenchman Hippolyte Fizeau (1819–1896), of a phenomenon that now carries both their names. The Doppler–Fizeau effect was originally purely acoustic. Every one of us experiences it daily. For this, it's enough to face a sound-emitting source that is moving, say, an ambulance siren. While the vehicle is approaching, the wailing of the siren is sharper than when it's stationary, but once it has passed, the wailing becomes flatter. This phenomenon is related to the wave nature of sound. Each note has its own wavelength. While the ambulance is not moving, I hear the original notes from the siren. But when the vehicle is moving and approaching me, the movement creates a

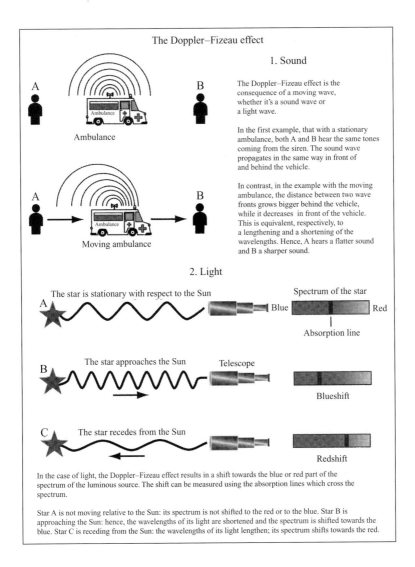

The Doppler–Fizeau effect

1. Sound

A

B

Ambulance

A

B

Moving ambulance

The Doppler–Fizeau effect is the consequence of a moving wave, whether it's a sound wave or a light wave.

In the first example, that with a stationary ambulance, both A and B hear the same tones coming from the siren. The sound wave propagates in the same way in front of and behind the vehicle.

In contrast, in the example with the moving ambulance, the distance between two wave fronts grows bigger behind the vehicle, while it decreases in front of the vehicle. This is equivalent, respectively, to a lengthening and a shortening of the wavelengths. Hence, A hears a flatter sound and B a sharper sound.

2. Light

The star is stationary with respect to the Sun

Spectrum of the star

A

Blue

Red

Absorption line

The star approaches the Sun

B

Telescope

Blueshift

The star recedes from the Sun

C

Redshift

In the case of light, the Doppler–Fizeau effect results in a shift towards the blue or red part of the spectrum of the luminous source. The shift can be measured using the absorption lines which cross the spectrum.

Star A is not moving relative to the Sun: its spectrum is not shifted to the red or to the blue. Star B is approaching the Sun: hence, the wavelengths of its light are shortened and the spectrum is shifted towards the blue. Star C is receding from the Sun: the wavelengths of its light lengthen; its spectrum shifts towards the red.

shortening of the wavelengths. A shortening of the wavelength means sharper notes, higher pitches. And vice versa.

Since light has a wavelike nature, the Doppler–Fizeau effect also applies to it, but with some significant differences, however. Rather than changing note, a light wave moving with respect to an observer changes colour. If the source is approaching, the wavelengths of its light shorten and shift towards the blue region of the spectrum. If it is receding, they shift towards the red, where the wavelengths are longer. You still need to measure these shifts. How can you do this? Thanks to the absorption lines, which are something like eyewitnesses. Huggins understood this. You just have to compare the positions of the same line, from the same chemical element, in the spectrum of a star and in the spectrum obtained in a laboratory, measure the difference between the two and then you can deduce the speed with which the star is approaching or receding.

The German Hermann Karl Vogel (1841–1907) refined the technique with the help of the Sun. As our star turns around an axis more or less parallel to that on which the Earth turns, its left edge approaches us while its right edge recedes (if North is to the top). Consequently, the light from the left edge should be blueshifted, while the light from the right edge should be redshifted. Vogel showed this effect with remarkable precision with the help of a technique that was in its infancy in the second half of the nineteenth century, photography. By fixing the light from spectra on a photosensitive chemical film, astrophysicists were able to measure spectral shifts at their leisure, with in addition a significant improvement in precision. From then on, spectrometry was transformed into spectrography.

At the close of the century, progress in spectrography made it possible to detect changes in speed of the order of ten, five and sometimes even three kilometres per second. This is easily enough not only to calculate stars' radial velocities, which are typically around 30 kilometres per second or so, but also possibly to detect the perturbations generated by star couples, at least for the brightest and closest among them.

A REVOLUTIONARY METHOD

After the Second World War, the spectrographic method underwent critical improvements that enabled it to attain hitherto unheard of accuracy. In 1953, the Briton Peter Felgett proposed, among other ideas, joining the spectrograph to a system that made it possible to measure the shifts of several spectral lines simultaneously. With such a setup, the precision should go from several tens of kilometres per second to a few hundred metres per second. The Briton's essentially theoretical idea was provocative, even revolutionary. This is why it was not until twenty years later that a researcher tried to implement Felgett's instrumental philosophy. It was another Briton, Roger Griffin, who had made the first 'cross correlation' spectrograph, which inspired us in Geneva and in Marseille to create machines like Coravel and Élodie.

Remember that the spectrum of a star is essentially a band of colours crossed by black lines. An essential detail for understanding our technique is that two stars with the same surface temperature (but not necessarily the same mass) globally produce the same spectrum and the same absorption lines. There is an intimate link between this surface temperature and the presence of such and such a chemical element in a stellar atmosphere. Felgett's idea was to take advantage of this rule to create a mask that reproduces a star's spectrum, but as a negative. Instead of being dark, the absorption lines are transparent, etched using a photochemical procedure on a piece of opaque glass. This mask is placed in front of the photodetector. You then just have to look at interesting stars, harvest their photons and transmit them through a series of lenses and refractive systems to extract the cleanest spectrum possible. You then gently slide it along until the absorption lines exactly face the troughs carved in the mask. You know that the two spectra face each other perfectly when the photodetector registers a minimum amount of light. You then just measure any possible spectral shift.

The first mask of the kind, Roger Griffin's one, had 240 slits spread over a section of the spectrum that was barely 500 Å wide (in comparison, the masks built for today's instruments have several thousand slits over a portion of the spectrum as wide as 2000–3000 Å). With this setup, the Briton's instrument shifted the field of spectrography into a higher gear. The efficiency gain with respect to its predecessors was a factor of 1000. And this was using improvements relating to precision, the numbers of lines measured and the speed of the measurements. Where before you needed half a night to get enough photons from a star to deduce its radial velocity, Griffin's system made it possible to carry out the same tasks in just a few minutes.

Our first spectrograph, Coravel, which started operation in 1977, was directly inspired by the Briton's instrument. With it, we attained a precision of 250 metres per second. At the time, we felt a bit like children spoilt with a magnificent present. Our instrument opened extraordinary research opportunities. We did, in fact, set off on an ambitious list of programmes: not all yielded results, but overall they were highly valuable, with that on the binary nature of stars in the Sun's neighbourhood being undoubtedly the most important for what was to follow. We did this for more than a decade.

During this time, other teams started on this new spectrographic path. For some of them, their goal was already to look for very low mass companions, brown dwarfs or possibly planets. But if they had high ambitions, the precision of their instruments had to be even higher. Jupiter, the most massive planet in the Solar System with its equivalent of 318 terrestrial masses, perturbs the Sun only very slightly. Expressed in radial velocity, this perturbation is only 13 metres per second. So it was crucial to improve the spectrographs.

THE FIRST PLANETARY SEARCHES

The Canadians Bruce Campbell and Gordon Walker, the very same who during the Baltimore conference would announce the discovery of very low mass objects, were pioneers in the high precision domain. In 1979, they published a fundamental article in which they revealed the principle of their instrument. Like us, they had been inspired by the work of Roger Griffin, and especially by an article written in 1973 in which the Briton explained how to improve the precision of a spectrograph by using the absorption lines of the terrestrial atmosphere. This method has a huge advantage: the reference spectrum and the stellar spectrum take the same path through the spectrograph, which reduces the risk of instrumental biases. Before entering a telescope, photons arriving from a star must first pass through our atmosphere, where they sometimes interact with the chemical elements there. In this way, the light collected by the spectrograph carries not only the absorption lines created in the star's atmosphere, but also those created in the Earth's atmosphere. Theoretically, it's enough to compare the two spectral signatures to establish any possible shifts in the light.

The principle is ingenious. In practice, things are different. To be really reliable, the absorption spectrum of the terrestrial atmosphere has to be very stable. But the atmosphere is not all stable. Just think of the winds that sweep through it nearly constantly and that push atoms along giving them some speed and hence some Doppler–Fizeau shift. The unfortunate result of this is that the weather can create the illusion that a star harbours a companion when it's not the case at all.

The Canadians kept only the best part of Griffin's idea and removed the drawbacks by enclosing a sort of artificial atmosphere in a 60-centimetre-long bulb which they placed at the telescope's focus. They used an atmosphere of hydrogen fluoride, a molecular gas, whose spectrum is characterised by absorption lines of exceptional clarity, which guarantees precision. The only drawback was that hydrogen fluoride is an especially poisonous, even lethal, gas and it's perfectly odourless. This is why many astronomers fearfully avoided approaching their Canadian colleagues' instrument which was installed on the 3.6-metre Canada–France–Hawaii telescope (CFHT) on Hawaii.

After several measurements carried out on the Sun, Campbell and Walker announced that they had attained the remarkable precision of 15 metres per second. Extrasolar jupiters were theoretically within their grasp. The quest could begin. And it took patience for the Canadians to see it through to the end. Jupiter makes a complete orbit around the Sun in just under 12 years. If all the gaseous giants in the Universe were like it, which is what the dominant theory of the time on planet formation imagined, then that would mean that astrophysicists would have to follow their stars for several years to hope to detect just a hint of a half-orbit, the minimum required to seriously consider the existence of an exoplanet. All of which doesn't help astronomers. Observing time on telescopes is limited. It has to be divided between dozens of teams who are generally carrying out very different programmes. Aware of the problem and of the time limits they had available, the two Canadians decided to retain a sample of only twenty stars.

Their strategy consisted of recording the spectrum of each star six times per year. In 1988, seven years after the beginning of this hunt, they were ready to announce the discovery of probable low mass companions around some of their stars. They were particularly interested in the behaviour of Gamma Cephei, an old orange star that shows a variation of 25 metres per second in amplitude. This could be due to an exoplanet located at some 300 million kilometres from the main star and endowed with an mass equal to one and a half that of Jupiter. Unfortunately for the Canadians, this proved not to be an exoplanet. In 1992, Gordon Walker was obliged to withdraw his initial conclusions after having discovered that the oscillation of 2.52 years was in fact due to the slow rotation of the star about itself.

AN UNCERTAIN MASS

HD 114762 B, the other candidate exoplanet presented at the Baltimore conference, also went through some pockets of turbulence. It's true that the two teams, that of Latham and our own, had carried out the same observation, but that wasn't enough to convince everyone. In 1992, two American researchers, William Cochran and Artie Hatzes, pointing out the slow rotation speed of the star, tried to show that HD 114762's companion was probably neither a planet nor a brown dwarf, but a more massive object, a small star in the red dwarf class.

The two University of Texas scientists had some legitimacy for their doubts. They too were in the game. They too were participating in the hunt for exoplanets, and had been since 1987 (a decade later, they codiscovered the planetary companion to the star 16 Cygnus B). Like the Canadians and like us in Geneva, they used the radial velocity spectrographic method. They could even boast of having attained high precision. They decided to include HD 114762 in their stellar sample

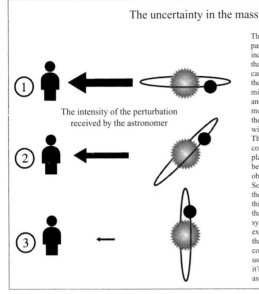

The uncertainty in the mass

The intensity of the perturbation received by the astronomer

The radial velocity method is particularly dependent on the inclination of the star–planet system that is being observed. As long as those carrying out the method don't know the inclination, they can only state the minimal mass of the object they detect and provide probabilities for how much greater it may be. In example 1, the system is exactly lined up with the astronomer's line of sight. The perturbation which is observed completely matches that which the planet induces in the star. In example 2, because the system is inclined, the observer only detects part of the effect. So, the estimated mass is lower than the reality. The astronomer considers this to be a minimal mass, knowing that it's subject to modification if the system's angle can be established. In example 3, the system is so inclined that the effect induced by the companion is no longer detectable using radial velocities. In contrast, it's the ideal configuration for astrometry.

just after Latham's announcement in Baltimore. They followed it for two years before making their criticism.

In the end, their work succumbed to later developments. But it's true that the debate would undoubtedly not have taken place if the radial velocity technique were able to accurately determine the mass of the stellar companions that it detects. All it can do is to make a minimum estimate and to calculate the probability that the mass is greater than the minimum. It owes this peculiarity to the fact that astronomers don't know at what angle we observe the systems that interest us.

Ideally, all exoplanets would orbit around their stars in planes that are perfectly aligned with our line of sight. If this were the case, then we would be sure that the perturbations observed in a star exactly correspond to the mass of the invisible companion. Alas, nothing is that simple. Extrasolar planetary systems can occur at any angle whatsoever. If a system is inclined by many degrees with respect to our line of sight, then part of the perturbation suffered by the star escapes from us, which yields a lower bound estimate to the companion's mass. This is why, in the absence of clear data on the inclination of an observed system, the spectrographic radial velocity method can only declare a minimum mass and an estimate of the probability of how close the mass of the object is to that minimum. In our equations, this uncertainty is completely accounted for in a factor, $\sin(i)$, which represents the inclination of the observed system and which forces us to resort to probabilities.

Getting back to HD 114762's companion, its minimum mass was established to be 11 Jupiters. This was well into the domain of exoplanets. However, these 11 jovian masses were very unlikely to correspond to reality. If the system had an average inclination, then the object's mass would approach 30–40 jovian masses. In the worst possible case, if the system is seen nearly face on, HD 114762 B could even be a small M dwarf, in other words a very faint star, but a star all the same. Personally, I don't believe that it could be a planet, and not just because of the probabilities. It was noticed that the star HD 114762 is particularly deficient in heavy elements, while all those stars around which planets have been discovered show strong metallicities (we'll get back to this later). So the brown dwarf hypothesis seems to me the most reasonable.

If this judgment is correct, then HD 114762 B is a member of the very tightknit club of brown dwarfs around Sun-like stars. Everything points to this configuration being particularly rare. Nature, for reasons that remain to be explained, seems to dislike great differences in mass. You can easily find high mass or low mass stellar couples, for example, binary brown dwarfs, but it's rare to come across couples made of two stars of which one is a heavyweight and the other is a lightweight. Why this scarcity? The question is far from being solved; indeed it's one of the major challenges for those who try to understand star formation.

Without knowing the reasons for this phenomenon, you can still try to observe its effects, like that which is now called the 'brown dwarf desert'. This is intimately linked to the distribution of companions of solar type stars. Up to about ten jovian masses, nearly all

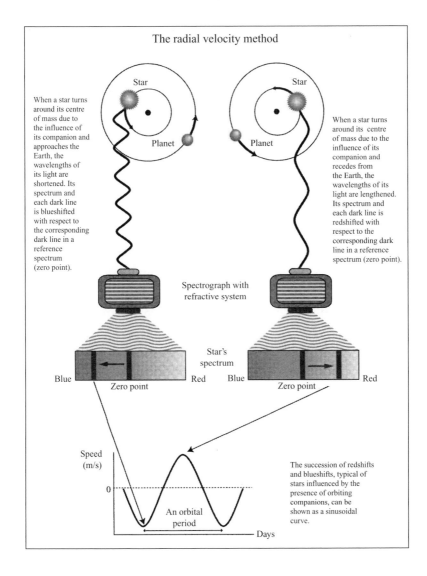

The radial velocity method

When a star turns around its centre of mass due to the influence of its companion and approaches the Earth, the wavelengths of its light are shortened. Its spectrum and each dark line is blueshifted with respect to the corresponding dark line in a reference spectrum (zero point).

Star

Planet

Star

Planet

When a star turns around its centre of mass due to the influence of its companion and recedes from the Earth, the wavelengths of its light are lengthened. Its spectrum and each dark line is redshifted with respect to the corresponding dark line in a reference spectrum (zero point).

Spectrograph with refractive system

Star's spectrum

Blue   Red   Blue   Red
Zero point   Zero point

Speed (m/s)

0

An orbital period

Days

The succession of redshifts and blueshifts, typical of stars influenced by the presence of orbiting companions, can be shown as a sinusoidal curve.

the sorts of exoplanets discovered are present. Then, starting from 13 jovian masses, begins the brown dwarf desert, which goes all the way up to about 80 jovian masses, the theoretical ceiling for failed stars. Let's be clear that it's not a completely dead and abandoned desert. But in the summer of 2000, you could only find four miserable brown dwarfs there. It is impossible to know if this is just an instrumental hangup. With the precision of today's spectrographs, we should be able to find them without any trouble if they were numerous. No, clearly, brown dwarfs refuse to be created in solar type environments, whereas they abound elsewhere (see the previous chapter).

Until now we've talked of several teams involved in the quest for exoplanets using the radial velocity method. We still have to mention one of the most important teams, our direct competitors, even though, as I would like to emphasise again, it's a very friendly and essentially positive competitiveness. When it happens that when we turn up at the same conference, each team brings wine from home and we drink these together and toast each others' discoveries.

Geoffrey Marcy, of the State University of San Francisco, has been interested in radial velocity since the beginning of the 1980s. He first became known for his work on brown dwarfs. He was then working in the medium precision domain, about 250 metres per second, with the spectrograph installed at the Lick Observatory telescope. His first study was on a sample of 70 red dwarfs (also called M dwarfs), which due to their low mass are more vulnerable to the gravitational influence of substellar mass companions. He also published an article devoted to the star Gliese 623, the companion of which was discovered by Sarah Lippincott, the American astrometric specialist. It had long been thought that this was a substellar mass object, maybe a big planet, or else undoubtedly a brown dwarf. However using his new, this time spectrographic, measurements, Marcy showed that Gliese's companion has a mass of 114 Jupiters, so it could well be a red dwarf rather than a brown dwarf.

Later, Marcy recruited the help of his friend and colleague Paul Butler, a chemist passionate about astronomy. Together, they put together a spectrograph able to measure high precision radial velocities. This was based on the instrument of the Canadians, Walker and Campbell, but Marcy and Butler refused to work with the infamous lethal gas, hydrogen fluoride. It's here that Paul Butler and his knowledge of chemistry turned out to be particularly valuable. His task was to find a molecular gas that had the same qualities as that used by the Canadians but without posing the same risk to the lives of the astronomers. Finally, he chose metallic iodine, vaporised and heated. This molecular gas is perfectly suitable. The first tests made on the Sun yielded a precision of 25 metres per second. This was a really excellent result. But the Marcy–Butler team was not satisfied. Their true aim was to go much lower, right down to 3 metres per second.

Thanks to the power of their instrument, the American team had a good chance of discovering very low mass objects. It could justifiably expect to find many. But instead of a miraculous harvest, Marcy announced to his colleagues, during a conference organised by the European Southern Observatory in 1994, that the search that he had carried out for two years had up till then failed utterly. There was not the merest hint of a substellar companion near any of the first 25 stars of his sample, which consisted of 75 stars in all. Among the experts, there was puzzlement. Could it be that exoplanets were so rare, that our Solar System was so unique?

It was at about this point that we inaugurated our new spectrograph Élodie and put it this time at the focus of the biggest telescope of the Observatoire de Haute-Provence, the 1.93-metre. Élodie had several improvements relative to its older sibling Coravel. It was born in the digital era. The new instrument was no longer equipped with a photodetector, but with a CCD array. The masking system on which the absorption lines were engraved in the negative was also abandoned for good in favour of several numerical masks managed by a computer, each corresponding to the spectrum of a particular type of star. Everything passed into the computers' intestines. It was they who calculated, compared, reduced and corrected in order to give us the only parameter that we cared about, the radial velocity.

THE FRANCO-SWISS SOLUTION

Like Coravel, Élodie was different from the spectrographs used by the Canadian and American teams. We didn't use a bulb filled with molecular gas which creates absorption lines. From the start, with our French colleagues, we preferred to apply another technique, that of an electric lamp containing a heated gas (plasma). This lamp, instead of creating dark absorption lines, produces emission lines that are characterised by their brightness and give us the zero points without which it would be impossible to precisely measure the spectral shifts.

In contrast to our colleagues from the other side of the Atlantic, we had to juggle with two sources of light: the reference lamp (that of Élodie contains thorium) and the star. As a result we increased the risk of instrumental bias. But the risk was worth it. First of all the risks were greatly reduced on Élodie by using optical fibres that conveyed the two light beams right up to the detector. Next, because this method allowed us to work over a larger portion of the spectrum than our colleagues could, we had at our disposal a larger number of absorption lines, which provides an extra handle on precision.

But it's possibly our software rather than the number of lines available that finally gave us the advantage. The solution used by Marcy and Butler at the time was a rather laborious one, and a rather slow one too. In contrast, our computer made it possible to obtain results just a few minutes after the observations at the telescope. This speed gave us the leisure to analyse the data while they were still hot, while our American colleagues were obliged to wait and archive their data before analysing them.

To this instrumental advantage, we added another, much more fortuitous one. While the American team, like the Canadian team, clearly planned to detect planets, we were mainly interested in objects like brown dwarfs. In contrast to the gaseous giants like Jupiter, no theoretical limit prevented brown dwarfs from occupying low – and therefore short – orbits around their main star. Even failed, brown dwarfs are stars, and we know from experience, thanks to previous observations, that some stellar couples have very short orbital periods, sometimes as short as just a few days.

So, in contrast to Marcy and Butler, we didn't concentrate on the typical perturbations of long orbits. This difference was to offer us the most beautiful of surprises, the discovery of a planetary companion whose presence made the radial velocity of its main star, 51 Peg, vary with an amplitude of 59 metres per second. With a period of 4.2 days, our object is a champion of short orbits, a record that becomes all the more significant when you know that the planet is just slightly less massive than Jupiter. This raised many a doubt on the reality of our discovery, since up till then no-one had thought that a gaseous giant could exist so close to its star. But a few days after the announcement of our discovery, our colleagues Marcy and Butler, who meanwhile had carried out their own measurements of 51 Peg, confirmed our data. Two months later, the same team announced in turn the discovery of two planets, 47 UMa b and 70 Vir b, the latter, despite a mass initially estimated at between 6 and 9 Jupiters, was also characterised by the relatively short orbital period of about 116 days.

A THREAT TO THE EXOPLANETS

Discoveries followed one after another throughout 1996, showing that the case of 51 Peg's companion was far from being exceptional. In spite of everything, some still doubted the reality of our strange planet. In February 1997, an article published in the review *Nature* questioned the existence of our object. It was signed by David Gray, an astronomer of the University of Western Ontario who was particularly interested in the oscillations that agitate the surfaces of stars. 51 Peg was part of his stellar sample. He had followed it since 1989 and had about forty measurements of it. He thought that the 4.2 day perturbation seen in 51 Peg was not due to the presence of a planet but to the regular pulsation of the star.

It's true that stars are not smooth and tranquil objects. Their surfaces are agitated, tremble, vibrate and reflect the bubbling which is born in the stellar core and which propagates all the way through to the upper layers of the stellar atmosphere. And then you need to take into account the contortions of the stars themselves. Since they are gaseous by nature, they don't rotate as solid bodies. The rotation accelerates as you get further from the equator. These differences cause the distortion of magnetic field lines which extend from one pole to the other. Forced by these contortions, the field lines end up forming knots, some of which cause the appearance of spots like those which you can regularly see on the surface of our Sun.

The possibility that these internal somersaults and the spots could be the cause of the oscillation detected in 51 Peg had not escaped us. These questions had indeed often been on our minds before we dared to announce our discovery. For example, we had rejected the influence of star spots after having seen that 51 Peg did not have the strong surface activity that generally accompanies the appearance of dark zones. Also, if the spots existed, then they would have to appear at regular intervals, every 4.2 days. Yet 51 Peg is a mature star, about 7 billion years old, which, like the other stars of its class, is characterised by a relatively slow rotation of about 30 days. So the spot hypothesis didn't stick. We also asked ourselves if a stellar pulsation could be the origin of 51 Peg's radial velocity oscillations. If that were the case, then we would have observed, by photometric techniques this time, characteristic luminosity and colour changes in 51 Peg. Yet our observations didn't show this.

But, in any case, our conclusions didn't convince David Gray, who favoured the existence – which he claimed with surprising vigour – of a 4.2 day stellar pulsation after having noted the distortions of absorption lines in 51 Peg's spectrum. I became aware of these conclusions in December 1996, just as I was leaving for a conference in Australia. Not being a specialist in stellar seismology, the science that deals with the ways in which stars shiver, I passed on the question to specialists that I found there. They were unanimous in saying that a 4.2 day pulsation was impossible. Nothing like it had been detected on the Sun, which is a very close cousin to 51 Peg. All the pulsations of our star seem to concentrate around a phase of 5 minutes. That invoked by David Gray is about 1200 times longer. The only possibility was that the pulsation had its origin in the gravity eigenmodes of the star, a special sort of wave that comes from the core of the star. But there too, the experts that I met in Australia were definite: gravity modes could not make it out to the surface; they don't pass the convective zone, one of the internal layers of the star.

Finally, very few scientists accepted David Gray's argument. Geoffrey Marcy, cited by an American magazine, declared, with his gift of the gab, that he was prepared to jump like an angel from the highest bridge in San Francisco if 51 Peg's oscillation was due to something other than a planetary companion. Two independent teams, that of Tim Brown and that of William Cochran and Artie Hatzes, in turn tried in vain to reproduce David Gray's measurements. Neither found anything to support the hypothesis of a 4.2 day pulsation.

Our planet resisted the scientific storm. There's now virtually no doubt regarding what it is. We'll just have to get used to its strangeness, as well as to that of the other exoplanets that up to now insist on defying the criteria that apply in our Solar System. It had long been thought that the Solar System reflected general laws that governed planet formation. It's clear that this is far from the truth. Theorists have no other choice than to go back to their drawing boards. They now have to explain the extraordinary variety of planets which occurs in the Universe.

8   Foreign planets different to our home-grown ones

Think of a scene rare on home-grown European TV, but common in the USA, which abounds with images of games of skill. Tenpin bowling is a highly refined art to those who are experts. In this sport, there's science in every step. From the calculation of the trajectory to the study of the resistance of the lanes, which are not uniformly rough. You can empathise with the fact that many years of skill are required to add the correct spin to the ball in order to attain, at the end of the lane, just the right curve which lets it hit the pins just slightly to the side in order that they fall like dominoes.

For beginners, the feat is infinitely more difficult. Either the ball, clearly wanting to be uncooperative, rushes off into the gulley, or else in spite of a nice, straight and apparently effective trajectory, it only removes the middle pins, leaving two separate groups of survivors, thereby removing all hope of cleaning out the set on the second throw.

So, frustration is often the lot of the novice, who is left with no option other than to persist in the hope that maybe one day. . . . Luckily, novices can count on an irregular ally: chance. Nothing seems to distinguish one bowl from the previous ones, yet miraculously the ball starts out in the lane, glides smoothly along before arriving at the rough bit, which it rapidly grasps, and perfectly curves its trajectory in just the way needed to leave no pin standing.

The planet 51 Pegasus b is a bit like the beginner's bowling ball which happened to make a strike, which shook all the pins of the dominant planet formation theory. Of course, it wouldn't have been able to cause all this trouble single-handedly. At first, it was just an oddball, a weirdo, an exception to the rule, a cosmic whim that was fun, an exotic circus animal that could astonish or even frighten you. But the number of such exceptions multiplied: Tau Bootis b,

55 Cancri b, Upsilon Andromedae b, three other examples among so many more of gaseous giants that cuddle up to their stars without disappearing in puffs of smoke.

Such promiscuity would have been less surprising if it had concerned a telluric planet like Mercury, which completes its orbit around the Sun in 84 days. But this was impossible. The instruments we had could never have detected such a low mass object. So it necessarily had to concern a colossus, weighing several hundreds of terrestrial masses. But back home in our Solar System, the giants are all far from the Sun. The first of them, Jupiter, with its 318 terrestrial masses, is at 5.2 astronomical units from its star, that is, 778 300 000 kilometres, and completes its orbit in 11 years and 315 days, while Neptune, which comes at the tail end of the march of the giants, has an average distance from the Sun of 4.5 billion kilometres.

This configuration emphatically put the experts who had worked on the planet formation theory in a tight spot. The surge of hot jupiters forced them to revise certain passages in their texts. And the more so because none of the exoplanets discovered up to now resembles our own. They're all different in one way or another, either in their distance from the star or the shape of their orbit. This makes us ask if our Solar System is not itself an exception. However, before being able to claim this, we'll have to discover many more planets and extrasolar systems.

PLANET FORMATION

Until the discovery of matter discs around certain stars (we'll get back to this later) and of exoplanets, the Solar System was the only possible observing ground for trying to understand how a star and its planet collection form. So it was from the Solar System that the dominant theory of planet formation was formulated. But let's start from the beginning.

The adventure starts with a great light and extreme heat, the Big Bang, which describes the birth not only of space and of time, but also of the matter that is spread throughout space and time. This matter is very largely composed of two essential elements, the lightest

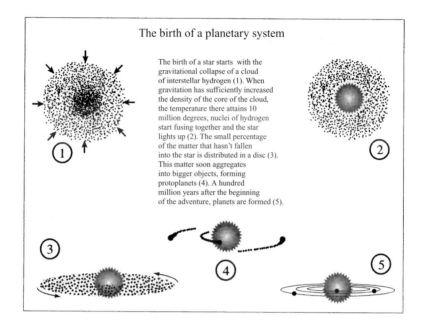

The birth of a planetary system

The birth of a star starts with the gravitational collapse of a cloud of interstellar hydrogen (1). When gravitation has sufficiently increased the density of the core of the cloud, the temperature there attains 10 million degrees, nuclei of hydrogen start fusing together and the star lights up (2). The small percentage of the matter that hasn't fallen into the star is distributed in a disc (3). This matter soon aggregates into bigger objects, forming protoplanets (4). A hundred million years after the beginning of the adventure, planets are formed (5).

which exist in the Universe: hydrogen (a proton and an electron) and helium (two protons, two neutrons and two electrons). These gases are spread throughout the space available, forming huge blobs of uneven density. They're littered with lumps, tracers of places where the gases concentrate the most. Gradually, encouraged by gravitation, which makes bodies attract each other and makes matter aggregate, these lumps move closer together, and finally collapse to higher and higher densities. In their cores, the pressure and temperature rise to incredibly high values. Nuclear fusion of hydrogen nuclei becomes possible. The first generation of stars has just lit up. They fill the Universe with their points of light. They're also the smiths of the cosmos, making the heavy elements, those that astronomers abusively call 'metals' (meaning everything that's heavier than the primordial gases, hydrogen and helium), in their hellish ovens.

These metals didn't exist before the birth of the first stars. The latter, and their descendants, produced the metals from the nuclear reactions in their cores and released them into space at the ends of their lives. This is how the Universe has been fertilised by heavy elements like iron, carbon, magnesium, silicon and, of course, oxygen. Scattered throughout the cosmos, these metals were mixed up in the clouds of helium and hydrogen and participated in the creation of new stars, more metallic than the previous ones.

Some 4.6 billion years ago – for the date, scientists have valuable clues from meteorites, some of which are the survivors of the proto-planetary epoch – one of these clouds collapsed on itself, constrained and forced by gravity. The Solar System was born from this unshapely lump, made up of 98% hydrogen and helium and 2% heavy elements. Our Sun used nearly all of the matter available. The rest, a tiny percentage, organised itself into a huge matter disc rotating around the Sun.

Before reaching adulthood, which allows it to enter into the main sequence (the club of stars that get their energy from hydrogen fusion), a star like the Sun passes through a particularly luminous phase. This phase contributes to heating up the matter that surrounds it, during which the protostar can dilate up to 100 astronomical units, i.e. 15 billion kilometres. Closer to the star, where temperatures of several hundreds of degrees dominate, only resistant compounds – called refractory material – can exist as solids. These are mainly silicates and metallic oxides, which constitute the primary material of the small interior, telluric, planets, like the Earth, Mars, Venus and Mercury.

Further from the Sun and its heat at about 4 or 5 astronomical units (750 million kilometres), the region where Jupiter orbits today, the temperature is much lower (about $-170\,^\circ\mathrm{C}$), allowing other elements to pass from a gaseous state to a solid state and form tiny grains. This is start of the ice kingdom, with ices made of ammonia, carbon monoxide and carbon dioxide, water and methane, which contribute to making the nuclei of the giant gaseous planets.

Metals, silicates, ices, a whole club of grains, neatly organised and shelved in the great rotating disc of matter, is in place, ready to deliver a great opus to posterity, a planetary opus. Except that the most difficult bit still has to be done. For the moment, the grains we're talking about, whether they're near the Sun or in the farflung suburbs, only reach infinitesimal proportions: a few microns (thousandths of a millimetre) at the most. How can dust turn into a planet? There are two theories.

The first invokes instabilities in the matter disc that is present around the star. This theory uses the same mechanisms that form stars. Instead of the gravitational collapse happening in a molecular cloud, it takes place in the disc of gas and dust constituted around the main star. Under the effect of their own gravitation, the regions in the disc that are a bit denser form lumps of solid matter. The gravitational collapse continues right up to planetary scales. The planetary gravitational collapse – and theorists are, for the moment, strict on this point – is only possible if the matter disc is relatively cold. If it's too hot, then a pressure typical for the gas succeeds in countering the process, and the lumps can't form.

Was this how Jupiter formed? Probably not. If it had been the case, its chemical composition would be relatively similar to that of the Sun. Yet many studies have shown that this giant gaseous planet is significantly richer in metallic elements than the Sun. This doesn't mean that in another planetary system, a jupiter can't form by gravitational collapse. But clearly, another mechanism is needed to explain how planets like it have formed in our Solar System.

The second theory, which largely dominates today, is historically the work of a Russian theorist, Viktor Safronov. It was then built up, thanks to the contributions of theorists like George Wetherhill, Stuart Weidenshilling and Jack Lissauer, to cite just a few. The scenario involves the phenomenon of accretion. Again we start with a disc made of gas and microscopic dust in rotation around its star. Everything turns around in the same general direction, though not necessarily straight ahead, nor at the same speed. Paths cross, aggregates form. The grains grow bigger. Soon they've become little pebbles that continue to collide. Sometimes they destroy each other, victims of relative velocities that are too high. But in the end, constructive

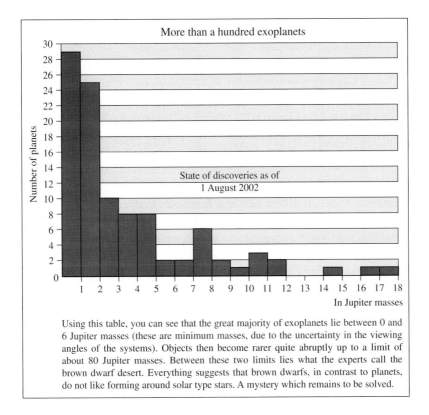

More than a hundred exoplanets

State of discoveries as of
1 August 2002

Number of planets

In Jupiter masses

Using this table, you can see that the great majority of exoplanets lie between 0 and
6 Jupiter masses (these are minimum masses, due to the uncertainty in the viewing
angles of the systems). Objects then become rarer quite abruptly up to a limit of
about 80 Jupiter masses. Between these two limits lies what the experts call the
brown dwarf desert. Everything suggests that brown dwarfs, in contrast to planets,
do not like forming around solar type stars. A mystery which remains to be solved.

collisions outweigh the destructive ones. Then the 'planetesimals' appear, in sizes ranging from metres to kilometres. They're the last step
before the making of planets. At first rather irregularly shaped, they
become more and more spherical due to the increase in their masses.

While this construction method can be used for all planets at
the beginnings of their lives, it doesn't always use the same materials.
Planets located close to the Sun, as we have seen, are mainly made
of metals and silicates. They're characterised by small masses partly
because of their short orbits, which allow less matter to be accumulated than long orbits. In short, the smaller the meadow, the less grass
there is to graze. In contrast, far from the Sun, the rings of matter are
not only bigger, but they're also richer thanks to the presence of the
ice grains. The gas giants such as Jupiter, Saturn, Uranus and Neptune

take advantage of this. Like the little planets they start their planetary careers with grain collisions. They become great big balls of dirty ice. They attain 7 or 8 terrestrial masses in just a few hundred thousand years. As matter becomes scarcer, their growth slows down. They take a few million years to attain the critical mass of 10–12 Earths. Then a new phase of rapid growth commences. The mass they build up produces a strong enough gravitational field to attract hydrogen and helium gas which is floating nearby. In a few hundred thousand years, depending on how much gas is available, the dirty ice planets become gaseous giants as massive as a few hundred Earths. And then, due to the lack of material, everything stops. The planetary heavyweights have gobbled up everything at their disposal, most probably clearing a corridor in the disc, a sort of empty ring or annulus along which they move from then on.

This planetary formation model, which favours collision between bigger and bigger grains, is undoubtedly that which is preferred by the majority of theorists. However, this doesn't stop it from having several grey zones. For example, the precise way in which the grains aggregate together is still under question. And the discovery of exoplanets brought up a whole new lot of puzzles.

In 1995, when Didier Queloz and I were rushing around trying to establish whether or not our discovery was a planet and whether it was viable since it was so close to its star, Alan Boss, of the Carnegie Institute in Washington, a respected theorist in planet formation, published an article in which he concluded that a gaseous giant was barely able to form at less than 4 or 5 astronomical units from its star, even if the latter is low mass and faint. This conclusion had a definite impact on the search for exoplanets. By stressing that gaseous giants were all far from their star, Alan Boss supported those using the astrometrical method, which is infinitely more sensitive to this sort of configuration than the radial velocity method, which by far favoured planets close to their stars.

Thus Alan Boss, like the great majority of the astronomical community, was surprised to learn that the first exoplanet was a gaseous giant stuck to its star and detected using the radial velocity method. What was this planet doing there, since theory, which seemed to be valid, said that it shouldn't be there? Maybe, said some, it was not a gaseous giant but a telluric giant, a heavyweight made of metallic oxides and silicates that formed on the spot. This is very unlikely. It is difficult to imagine that there could be enough matter so close to the star to form such a planetary sumo. No, you need to look elsewhere for an explanation.

THE MIGRATION OF GASEOUS GIANTS

We could have expected a long crossing of a theoretical desert, but instead, a very interesting idea quickly entered the field, just a few days after the anouncement of 51 Peg b's discovery. It came from three Americans, Douglas Lin, Peter Bodenheimer and Derek Richardson, who suggested the possibility of planetary migration. 51 Peg's companion could well have formed several astronomical units from its star, and then have moved towards it along a spiral trajectory, and suddenly stopped just at the threshold.

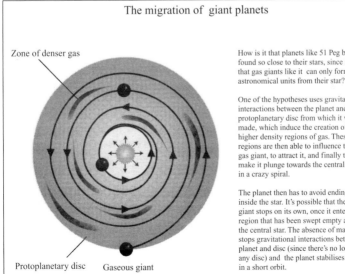

The migration of giant planets

Zone of denser gas

Protoplanetary disc   Gaseous giant

How is it that planets like 51 Peg b can be found so close to their stars, since it's known that gas giants like it can only form several astronomical units from their star?

One of the hypotheses uses gravitational interactions between the planet and the protoplanetary disc from which it was made, which induce the creation of higher density regions of gas. These regions are then able to influence the gas giant, to attract it, and finally to make it plunge towards the central star in a crazy spiral.

The planet then has to avoid ending up inside the star. It's possible that the gas giant stops on its own, once it enters the region that has been swept empty around the central star. The absence of matter stops gravitational interactions between planet and disc (since there's no longer any disc) and the planet stabilises in a short orbit.

Planetary migration is not a totally new theory. A small group of researchers had already considered about it for several years. Around 1980, two members of the California Institute of Technology, Peter Goldreich and Scott Tremaine, wondered about the strange links that exist between the rings and satellites of Saturn, which, according to them, could have caused migration of the satellites. Quite casually, they added that such a phenomenon was feasible for a young protostellar disc and forming planets. This theory was then taken up and developed by high flying scientists such as William Ward, Kerri Hourigan, Douglas Lin, John Papaloizou and Pavel Artymowicz.

The general scheme of migration is more or less the same from one author to another. A protoplanet forms in a young protostellar disc. As it grows, it increases in gravitational force. By doing this, it creates density waves in the disc. Where before the disc was relatively homogeneous, now long eyebrow-like structures appear: they mark places where the matter density is higher than elsewhere. Some of these paddings of excess matter are bigger than others. They happen to lie where the planetary motion is in resonance with the movements of the disc. What happens between the two is then what happens when the wind, because it's blowing in bursts of just the right frequency, induces the swinging to and fro of a suspension bridge weighing hundreds of tonnes as if it was just a ribbon of rubber.

Once the high density regions have formed in the disc, they in turn act on the planet by their gravitational attraction. As they exist on both sides of the planet, in the interior disc and in the exterior disc, the planet is pulled simultaneously in both directions. In the end, migration favours the interior direction, towards the star, whether or not the planet has emptied its corridor.

Now that we know how our gaseous giant has started its descent towards its star, all that remains is to imagine what could stop it. Because if it doesn't find a way of breaking its spiral infall, it'll end up being gobbled by the star. Lin, Bodenheimer and Richardson, and this is a major contribution of their article on the migration of 51 Peg b, presented two mechanisms capable of stopping a migrating jupiter early enough that it would avoid being eaten by its main star.

The first mechanism was inspired by a theory of the American Frank Shu, who described how a young star succeeds in creating a central cavity around itself. But in contrast to the big planets, which open corridors by pulling in surrounding matter, the star tidies its garden with the help of a magnetic field. The magnetic field, due to the rapid rotation of the young star, is impressively powerful. When ionised matter – in which the electrons have been removed from the atoms – comes in contact with it, it has no choice but to follow. It can only reach the star by the poles, where the field lines of the magnetic force enter and leave. The three Americans imagined that the planet falls gently down to this empty zone created in the equatorial plane of the star. As there's no more matter there, the planet can no longer create the density waves that induced its migration via gravitational interactions. So it stabilises in an orbit that corresponds to the place of corotation, where the planet and the star turn at the same speed.

In the second mechanism, the planet goes through its spiral migration towards the star, falling and slowing down, until it's given just enough extra orbital speed by the star's tidal forces in order to avoid falling further. It's a situation similar to that between the Moon and the Earth. Separated by a short distance, the two objects mutually distort one another until sort of outgrowths appear. Once the planet is close enough that it orbits slower than the star rotates, the star's outgrowth, because of its mass, gravitationally 'sticks' to the planet, giving it just enough extra speed to stop its fatal drop. The gaseous giant is finally stabilised in a low orbit.

A third scenario has to be added to these first two, that thought of by the Swiss Willy Benz and the American David Trilling. Once again, we start with a spiral migration induced by gravitational interactions between the disc and the gaseous giant. The latter plunges until it gets to the point we call the Roche limit, beyond which it would be destroyed by the tidal forces caused by the central star's attraction. Up to this point, the gravitational attraction between the two objects is defined by a figure of eight (rotated in three dimensions about its long axis), the two lobes of which are centred on the two objects (see the figure of the millisecond pulsar in Chapter 5). As the planet approaches the star, its lobe retracts, until the moment when it becomes smaller than the planet itself. From then on, the main star is capable of swallowing its companion – we've already seen this phenomenon at work in the formation of millisecond pulsars – and of sucking in its gas. Little by little, the planet loses matter, which forces it to distance itself from the star. But it doesn't go very far, because behind it, the matter disc continues to push. Sandwiched between the two forces, the planet stabilises in a low orbit.

So we have three scenarios to explain how a gas giant can migrate and then stabilise in a low orbit. It remains to be seen which is the right one, which can explain why 51 Peg b moved so close to its star without yielding to the burning melodies of the sirens. All bets are still open, but the odds are that each of these theories has some role to play. It's possible that they work together to explain the large variety of short orbital period gaseous giants, since 51 Peg b is not the only one to have these properties.

In the months that followed our discovery, our American competitors, Geoffrey Marcy and Paul Butler, announced the detection of several massive planets stuck to their central stars, such as Upsilon Andromeda b (0.73 Jupiter masses) and 55 Tau Boötes b (3.6 Jupiter masses). Today, we know of more than about fifteen planets that have orbital periods of less than 10 days. In August 2000, the record for the shortest orbit was held by a planet that we discovered around the star HD 83443 and which completes its orbit in 2.985 days (which, by the way, is not its only remarkable feature, since it's also one of the least massive planets known around ordinary stars, at only 0.35 Jupiters).

Another question in the migration model was important for the theorists, that of time scales. Many observations suggest that accretion discs, in other words those which are still made of dust and gas, disappear within ten million years, and often in less than five million years. To see this, it was necessary to observe very young stars that haven't yet finished contracting, during a phase called T Tauri, traversed by all stars that aim to join the main sequence. The first of these observations were made in the infrared. A satellite called IRAS (InfraRed Astronomical Satellite), the result of an American, Dutch and British collaboration, contributed a lot to knowledge in this area. Using it, it was noticed, in particular, that young stars show excesses of infrared radiation. This is a very special signature that can only be the product of a young disc of gas and dust.

So less than ten million years is the time available for a planet like Jupiter to form. After this period, the disc has lost virtually all of its hydrogen and helium. Either the gas has fallen onto the star, nourishing it further, or else it has been thrown out into space by the very intense activity characterising the childhood of the star. If a giant protoplanet has the ambition of cloaking itself in a thick, gaseous atmosphere, it's in its best interest to do this as soon as possible.

The telluric planets don't have the same sense of urgency. Their primary matter, made of stones, better resists the whims of the star. While Jupiter had just ten million years in which to form, we estimate that the Earth arrived at its final mass after a hundred million years. It's a weird Universe in which the more massive a planet is, the less time it has to make itself. You can imagine that in many developing planetary systems, the gas in the accretion disc disappears too quickly for gas giants to form.

It's also possible that gas giants can form, but that the disc disappears immediately afterwards, preventing the migration phase. Maybe this is what happened during the childhood of our Solar System and allowed our giant planets to stay put in the orbits where they were born. If this hadn't been the case, we would certainly not have been here to talk about it. A jupiter plunging towards the Sun could not have failed to influence the orbit of the Earth to such an extent that the Earth would have certainly been either thrown into the Sun or thrown out into interstellar space. From this argument to saying that there's a relation between the longevity of accretion discs and the frequency of telluric planets is a step that many experts take quite reasonably.

Other theorists, such as Frederic Rasio and Eric Ford, of MIT, don't have this problem of a race against time. In their migration model, inspired from the work of theorists like Stuart Weidenschilling of the University of Arizona and Francesco Marzari of the University of Padova, there's no need for an accretion disc for this mode of travel. Sure, gravitational interactions still come into play, but this time they happen between several gas giants, members of the same planetary system. If each planet were a Kalashnikov-toting warlord, you would hear him say 'This town is too small for all of us. Some of us have to go', since too many massive giants in too tight a space makes it nearly impossible for all of them to have stable orbits. So a succession of gravitational interactions commences which can lead to the ejection of one giant into interstellar space and the migration of the other closer to the star.

VERY ECCENTRIC ORBITS

This game of mutual influence certainly throws a different light onto the causes of planetary migration. But it does much more besides. It may hold the key to another mystery, that of the orbital eccentricity of exoplanets, at least of those that are relatively far from their stars. For the others, the hot jupiters, circular orbits are usual, undoubtedly induced by long migrations which had the effect of removing any orbital eccentricities.

In our own Solar System, planets tend to follow orbits that are more circular than eccentric. A fact that theories before 1996 explained by emphasising the capacity of protoplanetary discs to circularise the orbits of objects forming in them. Unfortunately, the exoplanets couldn't care less about the old models! Take HD 80606's companion, for example, which we discovered with our colleague David Latham and whose eccentricity rewrote the record books since its numerical value is 0.927. To calculate an eccentricity, you divide the smallest planet–star distance (the periastron) by half the length of the long axis of the ellipse and you subtract this from 1. An eccentricity of 0 corresponds to a perfect circle and as you increase

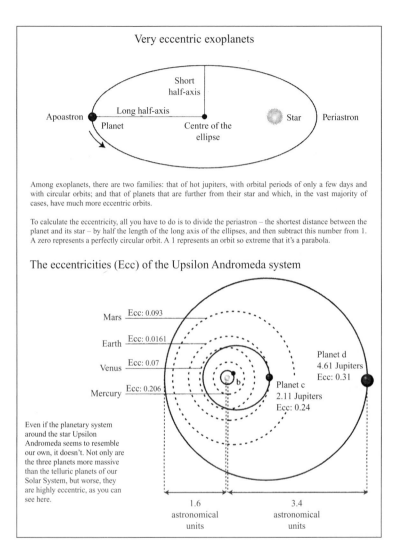

Very eccentric exoplanets

Among exoplanets, there are two families: that of hot jupiters, with orbital periods of only a few days and with circular orbits; and that of planets that are further from their star and which, in the vast majority of cases, have much more eccentric orbits.

To calculate the eccentricity, all you have to do is to divide the periastron – the shortest distance between the planet and its star – by half the length of the long axis of the ellipses, and then subtract this number from 1. A zero represents a perfectly circular orbit. A 1 represents an orbit so extreme that it's a parabola.

The eccentricities (Ecc) of the Upsilon Andromeda system

Mars — Ecc: 0.093

Earth — Ecc: 0.0161

Venus — Ecc: 0.07

Mercury — Ecc: 0.206

Planet d
4.61 Jupiters
Ecc: 0.31

Planet c
2.11 Jupiters
Ecc: 0.24

b

Even if the planetary system around the star Upsilon Andromeda seems to resemble our own, it doesn't. Not only are the three planets more massive than the telluric planets of our Solar System, but worse, they are highly eccentric, as you can see here.

1.6
astronomical
units

3.4
astronomical
units

the eccentricity towards 1, the ellipse becomes longer and in the limit becomes a parabola. Above 1, you have a hyperbola. In contrast, Jupiter's eccentricity is only 0.048. This is typical of planets in the Solar System; it's almost perfectly circular. So, why has HD 80606 b, this heavyweight of about four jovian masses with a period of 112 days, adopted such an elliptical orbit? It's a mystery. And the question holds for many objects discovered since 1996.

In summary, we have two families of exoplanets. On the one hand, there are the hot gas giants of the 51 Peg b type, which are cuddled up to their stars (at less than 0.2 astronomical units), usually light, less than a Jupiter mass (except for Tau Boötes), and characterised by circular orbits. On the other hand, there are the planets which lie further out (between 0.2 and 3 astronomical units), and which are on average heavier, above a jovian mass (but again there are exceptions), and generally characterised by eccentric orbits. While the former could have followed orthodox migrations, the latter could be the survivors of a complex game of gravitational influence between several gas giants.

But this neat picture is far from convincing everyone. In 1998, Pavel Artymowicz, from the Observatory of Stockholm, and known for having proposed a mechanism that enables gas giants to continue to attract matter even when they've created a corridor (and so can surpass a Jupiter mass), developed the idea that the more massive a planet is, the more its interactions with the disc would push it into an eccentric orbit.

While his model works well for average eccentricities, it has difficulty explaining extreme cases, like that of the companion of 16 Cygnus B, which was jointly discovered in 1996 by William Cochran and Artie Hatzes and by Geoffrey Marcy and Paul Butler. This planet has 1.6 times Jupiter's mass and an eccentricity of 0.6. A high eccentricity and a low mass, it's enough to send anyone mad, unless the ellipse is the result of chaotic gravitational influences.

The system 16 Cygnus has three stars. A and B, separated at the closest by about 1000 astronomical units, orbit around each other in 125 000 years, while C, a red dwarf, prefers to stay 100 000 astronomical units away from its two sisters. The companion of 16 Cygnus B grew up in this family. It had nothing to complain about with its aunt C, which was much too far away to influence its education. On the other hand, aunt A started interfering when the planet was a toddler. And so the orbit of the poor planetary infant, torn between its two parents, A and B, was stretched more and more until it ended up with this record eccentricity. But nothing says that it can't have also been subjected to chaotic interactions induced by the presence of other planets. Who knows? We might find other objects around 16 Cygni B some day. We would then have a multiple exoplanet system.

FINALLY, SYSTEMS WITH MULTIPLE PLANETS

The verdict was already known for Upsilon Andromeda, the site of the first multi-planet system detected around a main sequence star. Jointly announced in 1999 by two independent teams, that of Geoffrey Marcy and Debra Fisher of the State University of San Francisco, associated with Paul Butler, of the Anglo-Australian Observatory, and that of Robert Noyes of the Harvard–Smithsonian Center for Astrophysics, and Timothy Brown, of the High Altitude Observatory of Boulder, this discovery was a crucial step in the quest for exoplanets.

The story started in June 1996, when the friends Marcy and Butler delivered their third batch of planets for the year. In their basket, there was a 0.8 jovian mass companion taking 4.6 days to orbit a star 44 light-years from the Sun, Upsilon Andromeda. Following this, it took nearly three years for the two teams to analyse the residuals of the first companion's curve, which it was thought was hiding the signatures of additional planets. As the months went by, the researchers continued to measure the star's perturbations in order to obtain data that were as precise as possible and to eliminate all possible sources of error. Soon, there was no longer any doubt. Upsilon Andromeda sheltered not one but three planets. We know the first, a hot jupiter. The second, at 0.82 astronomical units (expressed as an average, a mean), weighs 2 Jupiters. While the third, at 2.4 astronomical units, weights about 4 Jupiters.

Like their extrasolar sisters of the same family, the two exterior planets of Upsilon Andromeda are characterised by prominent eccentricities, 0.23 and 0.31 respectively. Having planets with high masses, that are close to the star and eccentric, the Upsilon Andromeda system is definitely nothing like our own. This led Geoffrey Marcy to declare to the *New York Times* (16 April 1999): *'I'm really curious*

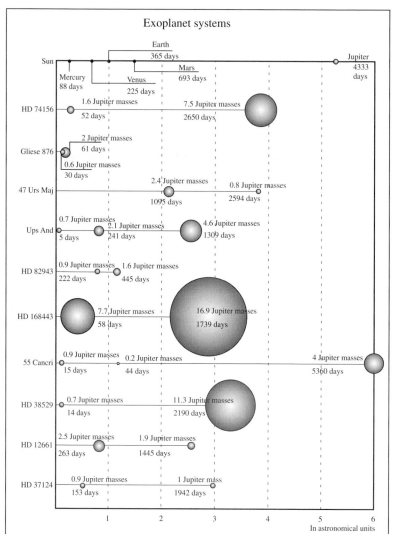

Exoplanet systems

By August 2002, ten exoplanetary systems had been discovered. You can see on this diagram that none of them really look like the Solar System. Is the Solar System unusual? To really know, we'll have to wait for instruments that are able to detect planets with masses as low as that of the Earth. (The masses are given here in units of one Jupiter mass and the orbital periods in terrestrial days. One astronomical unit equals about 150 million kilometres.)

to know how such a system could have got into place. This will no doubt contribute to shaking up the standard planetary formation theories.... We can continue to wonder whether or not our Solar System is unique in some way.'

At the Observatory of Geneva, we had long thought we had found a second multiplanetary extrasolar system. Here again, this happened in several steps. Early in the year 2000, we discovered a first planet with the mass of Saturn orbiting in just 2.985 days around HD 83443, a star at the distance of 141 light-years from the Sun. Then, after having analysed the residuals from that first oscillation, we noticed a second oscillation that could have been that of a second planet with a 29.8 day period and with a quite humble mass, about half that of Saturn. We published all the data, but continued to follow HD 83443. However, to our great suprise, the measurements that we gathered during 2001 and 2002 showed no further trace of the second oscillation. We have to admit that something internal to the star, maybe a significant group of sunspots, led us astray. Luckily, we had better luck with other stars.

By early August 2002, at least ten extrasolar systems had been discovered: Gliese 876, 47 Ursae Majoris, Upsilon Andromedae, HD 82943, HD 168443, HD 74156, 55 Cancri (HD 75732), HD 38529, HD 12661 and HD 37124. A marvellous harvest. These systems are nothing like our own, the Solar System, though all yield precious information. In the case of Gliese 876 and HD 82943, we have resonant systems. The orbital periods of their respective planets follow whole number proportions, and in fact it's a factor of 2 in both cases: while the outer planet completes one orbit, her sister, closer to the star, has time to go around twice. Such configurations support the migration theory. You can easily imagine planets sliding through the accretion disc, gravitationally sticking, influencing one another, until the point at which they've stabilised their orbits in resonance. But these are only hypotheses and we need many more such systems in order to construct a solid theory of how they were made.

It would be best if we could watch each step of the birth of a planetary system while it's happening. A volunteer is needed to sit at a powerful telescope, to gaze at one of the bubbles of gas and dust that is forming, for example, in the region of the Orion Nebula – of which the Hubble telescope has made some magnificent images – and to patiently follow it, watch it collapse, light up its star, flatten its disc, see the grains condense in a way that depends on their distance from the star and on their chemical composition, see planetesimals form, planets being made and maybe migrating, or maybe playing gravitational billiards, and wait for the survivors to stabilise. However, this is a research project that would be long term, nearly a hundred million years, if the formation of the Earth is typical. So, to fill in the holes in their knowledge, astronomers rely on different methods, for example the discovery of other solar systems and the classification of these.

But they also rely on searching through the sky looking for discs of different ages, seen at different stages of their evolution – accretion, protoplanet, debris – in order to reconstruct the film of planetary formation. These researchers are like beings that have lifetimes of less than a second but who would like to know what a human being is. Not being able to follow a single individual during his whole lifetime, the only option would be to photograph large numbers of men and women, among whom you would find babies, children, adolescents, adults and senior citizens, and to use this crowd to reconstruct the development of the human being.

THE DISCOVERY OF OTHER DISCS

As we said above, very young discs of gas and dust have been discovered around T Tauri stars, aged less than 10 million years, which haven't yet finished contracting. These first detections, which we owe to the IRAS satellite, were carried out in the infrared.

Instruments have evolved since the 1980s and now photographs of these very young objects are available. There are even magnificent ones that the Hubble telescope has sent us from one of its optical voyages in the Orion Nebula, known as a stellar breeding ground. In the images, you can distinguish what looks like a dark eye, the disc, with a bright pupil at the centre, the star. Surveys have been carried out to establish the proportion of stars that are surrounded by such discs: the survey by the American Ray Jayawardhana in the stellar association TW Hya has shown that only stars younger than 10 million years show accretion discs.

In the others, dust has already coalesced into planetesimals and most of the gas has disappeared. The star HR 4796 A is just over ten million years old and its disc is apparently undergoing rapid change. In 1998, some American researchers showed that it has a large interior void, a space as large as our Solar System, in other words about 30 astronomical units. Does this mean that some gas giants have cleaned out the interior and that some small telluric planets are starting their long period of growth? Who knows . . . ?

From the youngest stars, let's move on to the oldest ones. It's the European satellite ISO (Infrared Space Observatory) that is credited with one of the records for discovering old discs when in 1998 it found an object spread around 55 Cancer, a star already known to harbouring one exoplanet and undoubtedly a second. The future will tell. As 55 Cancer's age is calculated to be about 5 billion years, that's also the age of its disc. It's out of the question for the latter to play a role in the accretion process of the star or of planets. It's only debris, a bit like that which exists in our Solar System: the asteroid belt between Mars and Jupiter, the Kuiper Belt, of which Pluto is the biggest member, and the Oort Cloud, located much further out, a reservoir of long-period comets.

Comets are among the oldest eyewitnesses of protoplanetary discs, in any case in our Solar System. This is why scientists are trying to determine their composition better and are sending probes like Giotto (whose objective was Halley's comet) into their paths in order to delve into their hearts and tails. But do they only exist in our Solar System? No, and we learned this from the IRAS satellite, which for ten months in 1983, passed the sky through a fine-tooth comb with its infrared detectors and brought back nearly 200 000 interesting shots.

One of the major contributions is undoubtedly the discovery of a strange object around Vega, the fifth brightest star in the sky. At 400 million years old – our Sun has notched about 5 billion springtimes – Vega was only meant to be a testbed to check that the satellite was working properly. It was expected that it would have an infrared signature worthy of a star of its class. Instead, astronomers noted a relatively large infrared excess. Was IRAS confused? That was the first thought that crossed people's minds, but all other measurements were found to be correct. So there had to be something unusual about Vega. How could the infrared signature be explained? Sure, it would be normal for a cold object with a temperature of about −180 °C, but not for a star bubbling with life and full of energy. So what caused it? Only one explanation remained: around bright Vega, there exists a matter disc that extends for several astronomical units and that, lightly caressed by stellar light rays, emits the infrared excess.

Again thanks to IRAS, three other stars soon revealed identical signatures: Fomalhaut, Epsilon Eridanus (for which the discovery of a companion was announced in the summer of 2000) and Beta Pictoris. In April 1984, while they were in Chile to look at Uranus and Neptune, Bradford Smith of the University of Arizona and Richard Terrile of the Jet Propulsion Laboratory decided to point their telescope at the stars hitting the headlines. They were then working on a coronagraph, that made it possible to mask the target star in order to observe its surroundings. By this means it was hoped that visible traces of discs would be revealed.

What they discovered left them breathless. There, on their screen, the masking of Beta Pictoris allowed a magnificent disc to appear in visible light, in profile, right in our line of sight. According to the first estimates, it extends up to 400 astronomical units (today, it is thought to be more like 800 or even 1000 astronomical units). The object very quickly became a much studied subject. The French astronomers Alfred Vidal-Madjar, Anne-Marie Lagrange and Roger Ferlet obtained the most crucial results. A spectroscopic study revealed frequent jets of hot gas in the neighbourhood of the star. According to the French, these comets suddenly evaporated as they plunged towards the main star. This would also explain the signature – rare – of carbon monoxide in the disc's spectrum.

At 100 million years old, as old as its star, the disc is probably a member of the class of debris discs. A hundred million years was the time needed for the Earth to attain its final mass. But even then it still had to cope with numerous meteoritic impacts. The conclusion is that if it resembles our Solar System, the Beta Pictoris system is getting its finishing touches. So, if you dared, you could just about imagine planets wandering around there....

Rather than imagining, some prefer to argue. So Vidal-Madjar and his colleagues refer to the presence of a planet in order to explain the cometary downpour on the star Beta Pictoris. The orbit of the planet – in fact, the model works better with two planets – would cross or approach close to a great ring of comets, destabilising the latter and pushing them into one-way journeys.

Another Frenchman added new clues in 1995. Alain Lecavelier des Étangs, also from the Institut d'Astrophysique de Paris, found some astonishing data on the star Beta Pictoris in the astronomical archives of the small photometric telescope of the Observatory of Geneva at La Silla (Chile). On 10 November 1981, the star had an abrupt and temporary drop of about 4% in its luminosity. Possibly a big comet and its dust tail had dulled its brightness by passing in front. Unless it was a planet.

Many other signs suggested a planetary presence near Beta Pictoris. In particular, telescopes had indicated a gap around the star, a space without dust that was as large as our Solar System. It's possible that several planets are criss-crossing the gap. There are also asymmetries in the disc, big asymmetries, in luminosity and in depth, imbalances that could be due to a massive planet whose orbit is offset by a few degrees from the average plane of the disc.

LOOKING AT METALLIC STARS

The stars themselves can provide valuable clues to the occurrence of planets. These are chemical clues like the amount of metallicity.

Guillermo Gonzalez, a researcher at the University of Washington, is less interested in exoplanets than in their respective suns, whose spectra he studies in order to determine their abundances of heavy elements, the metals that we've already mentioned in talking about how planets form. His first study was in 1997, and the results were, to say the least, surprising: four of the five stars whose planets occur at less than 0.3 astronomical units show a noticeably higher metallicity than average. The sample was too small to draw any serious conclusions, but the path taken by Guillermo Gonzalez was certainly an interesting one, even if very few believed in his work at the time.

In 1999, he published new results, this time based on the study of a dozen stars. His conclusions hadn't changed. The stars accompanied by gas giants are really richer in heavy elements than average stars. Some of them even flirt with records, starting with 55 Cancer (yet again!) and 14 Hercules, which are among the Sun's most metallic neighbours.

Theorists put forward two explanations for this strange correlation. Either the abundance of heavy elements present in a protostellar disc helps in the construction of planets around a central star, or else it's the construction of planets in a disc and later the collision of some of these with the star that increases the heavy element abundance of the latter.

It's difficult to decide which. Unless.... Take the case of HD 82943, a star accompanied by at least two giant planets, which we discovered in 2001 with the help of our colleagues Rafel Rebolo and Garik Israelian from the Canaries Observatory. On painstakingly studying its spectrum, we noticed some unusual signs, but we had to confirm the results using a more powerful instrument than Coralie. So, we redid the measurements with UVES, the high resolution spectrograph mounted at the focus of one of the 8.2-metre telescopes of the VLT, in Chile. Our first impression had been right. There's lithium-6 in the atmosphere of the star HD 82943.

We first came across the use of lithium spectral lines in the identification of brown dwarfs. Here, the technique was identical, except that there was a difference in isotopes. Lithium-7 is different to lithium-6. Both have the same numbers of protons and electrons, but lithium-6 has one neutron less than its cousin. A difference of just one particle might seem rather trivial, but it leads to significant consequences. While lithium-7 can exist in a stellar atmosphere, lithium-6 rapidly disappears early in the youth of a star. In other words, if we can see the trace in an evolved star, this isotope can't come from the original material from which the star was made. Only one explanation remains: the lithium-6 must come from one or several planets, which, as victims of unstable orbits, have plunged towards their sun, dispersing their elements in its outer layers. In the case of HD 82943, the amount detected makes us think that a giant planet twice as massive as Jupiter must have been swallowed. Unless there were three telluric planets of the Earth's mass.

There's no doubt that the metallicity is a good indicator of the presence of planets around a star. After all, it was by searching for especially metallic stars that, at the beginning of 2000, Geoffrey Marcy discovered the star BD 103166 and, orbiting around it, a gas giant of 0.48 jovian masses and an orbital period of about 3.5 days.

As for us, we plan to continue using the radial velocity spectrographic method. This method has already allowed us to detect the perturbations induced by planets less massive than Saturn, such as HD 83343 c. We hope to do even better yet with our next instrument. Called Harps, it was delivered to the large 3.6-metre telescope in La Silla, Chile, in December 2002. In early 2003, we went to Chile to commission Harps and all the calibration tests gave excellent results. We expect it to attain a precision of 1 metre per second. That should allow us to detect not only giant planets further still from their stars, but also lighter planets, maybe even of a few terrestrial masses, if it happens that they have orbital periods of less than 10 days.

But with a precision of 1 metre per second, the radial velocity method will reach its limit. Sure, it's theoretically possible to detect variations of the order of a centimetre per second. But this can only be done with very close objects. Once distances are in the light-year range, its vision becomes fuzzy and it's difficult to say whether perturbations of less than 1 metre per second are due to a companion or the star itself. You just have to think of the Sun and its regular pulsations every five minutes: stars shiver, cough, dance, sneeze. Even the arrival of large sunspots can influence a spectrograph.

While it has been effective for detecting extrasolar gas giants, the radial velocity method therefore needs to give way to other techniques, since the aim is to detect planets like the Earth, whose perturbation on the Sun, expressed as a radial velocity, is not more than 8 centimetres per second, compared to 13 metres per second for Jupiter.

As satellites become more and more powerful, astrometry, reinforced by techniques such as interferometry, will reappear phoenix-like from its ashes. But science wants more. In the end, it would like to see planets in the infrared or in visible light. It even dreams of the coloured contours of a far-off Earth, with the white of its clouds, the blue of its oceans and the brown of its continents.

nature

INTERNATIONAL WEEKLY JOURNAL OF SCIENCE

Volume 378   No. 6555   23 November 1995

A planet in Pegasus?

Stochastic resonance in ion channels

The becoming of birds

NO behavioural role

1. In November 1995, the scientific review *Nature* cautiously announced, with a question mark, the discovery of the first exoplanet around a main sequence star, the star 51 Peg. Seven years later, exoplanet hunters have detected more than a hundred stars.

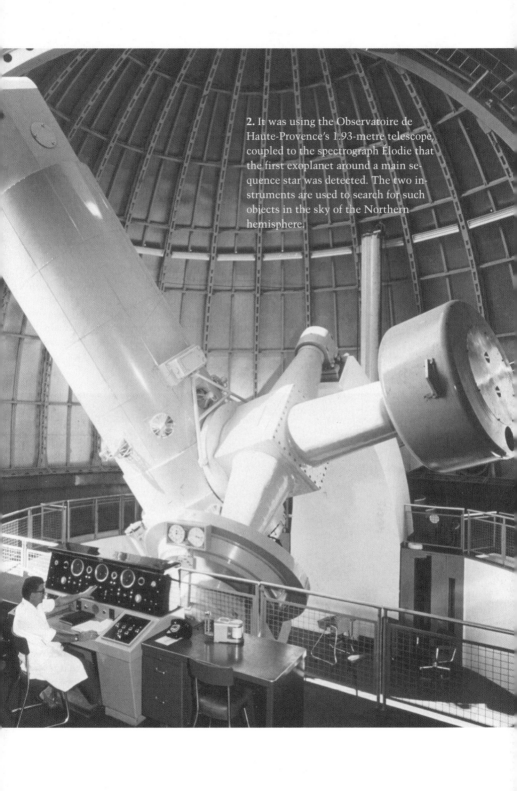

2. It was using the Observatoire de Haute-Provence's 1.93-metre telescope, coupled to the spectrograph Élodie that the first exoplanet around a main sequence star was detected. The two instruments are used to search for such objects in the sky of the Northern hemisphere.

3. The Very Large Telescope (VLT), the pride of the European Southern Observatory (ESO), is located at Paranal, in northern Chile. Right now the finishing touches are being put into place. There are four giant 8.2-metre diameter mirrors. Only the Keck, an American instrument installed in Hawaii, with its two 10-metre mirrors, can compete with its power. The VLT will undoubtedly enable significant work to be done in the exoplanet domain.

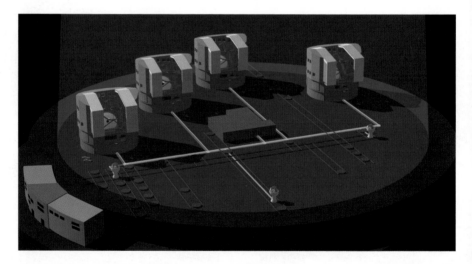

4. Three small 1.8-metre mobile telescopes will soon be added to the four giant mirrors at the VLT. By combining the light received by the small mobile telescopes with that received by one or more large mirrors, according to the principles of interferometry, it'll be possible to obtain the same resolution as a giant telescope several dozens of metres in diameter. Here again, this power will be useful for work concerning exoplanets.

interstellar gas clouds and cause some of them to collapse and finally form new stars. This is why the spiral arms always contain very young, very bright stars.

6. Photographed by the Hubble Space Telescope, these huge interstellar gas columns are located in the Eagle Nebula, also called M16. Experts see a future stellar nursery in this cosmic sculpture lit up by a few close stars. A few gaseous blobs will gently break off, collapse on themselves and turn into stars..

gions), from where earths and jupiters might emerge.

8. The dominant theory of planet formation invokes a phenomenon in which dust and ices form bigger and bigger chunks until they become planets. In this simulation, carried out by a researcher from the Lick Observatory, you can see the empty ring and distortions created in the protoplanetary disc by a gas giant that is forming. Interactions between the disc and the planet would seem to explain why some gas giants migrate towards their stars.

9. Beta Pictoris is the first star whose matter disc was photographed. It's seen perfectly in profile with respect to our line of sight. It's not a protoplanetary disc, but rather a debris disc, since the central star is several hundred million years old. If there are any planets, they will have already formed. In fact, a deformation in the disc, which can be seen in these two photos, leads researchers to think that there may be a planet around β Pictoris.

Size of Pluto's orbit

10. Science fiction made Mars the planet of little green men eager to conquer the Earth. In reality, it's the opposite that is more likely to happen. The red planet is, in fact, at the top of the list for the American and European space agencies for looking for possible traces of life.

440 yd
400 m

11. Liquid water, which is essential to life, may have existed over large areas of Mars, many hundreds of millions of years ago. What remains today? On this image taken by the American probe Mars Global Surveyor in June 2000, you can see, in particular, ravines that cross the sides of an ancient crater. According to some, this erosion is due to frequent surges of water, which could maybe still occur.

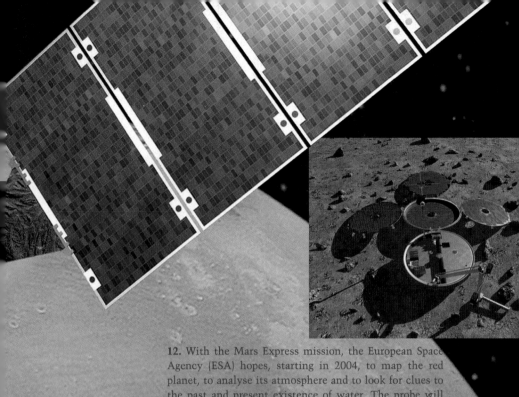

12. With the Mars Express mission, the European Space Agency (ESA) hopes, starting in 2004, to map the red planet, to analyse its atmosphere and to look for clues to the past and present existence of water. The probe will also release a small permanent craft, named Beagle 2, mainly constructed by British researchers, which should analyse the ground using a mass spectrometer, that is able to measure certain products of biochemical activity.

13. On the other side of the Atlantic, at NASA, a Martian mission for 2003 is being planned around two mobile 150-kilogram robots. While these robots will not have instruments as sensitive as the Beagle 2 spectrometer, they'll have two advantages: they will be able to move about 100 metres per day in order to carry out a large scale geological exploration, and as there are two of them, this reduces the risk of failure.

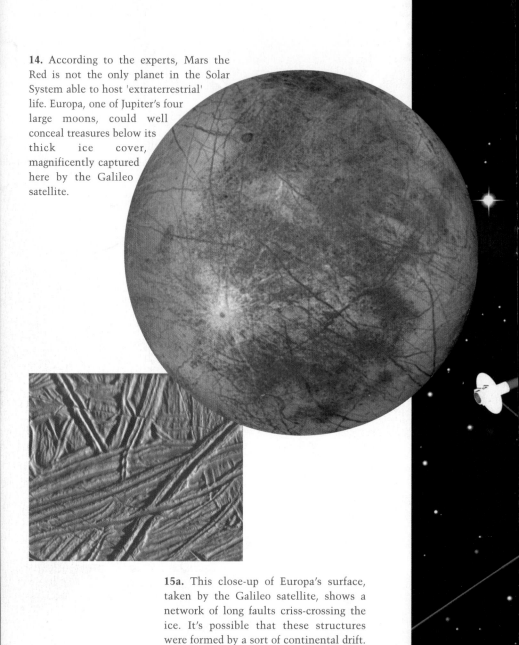

14. According to the experts, Mars the Red is not the only planet in the Solar System able to host 'extraterrestrial' life. Europa, one of Jupiter's four large moons, could well conceal treasures below its thick ice cover, magnificently captured here by the Galileo satellite.

15a. This close-up of Europa's surface, taken by the Galileo satellite, shows a network of long faults criss-crossing the ice. It's possible that these structures were formed by a sort of continental drift. If this is the case, then it's also possible that this thick ice layer is sliding across a water base that has been liquefied by deep volcanic activity. With water and a source of heat and energy, it may have been possible for single-celled life to develop.

15b. The search for extraterrestrial life is not confined to the Solar System. Astrophysicists already plan to survey the atmospheres of distant planets several light-years away to try to detect signs of life, like ozone, a molecule made of three oxygen atoms. This will be possible using a new generation of telescopes, space telescopes composed of several mirrors 'flying' in formation and working according to the principles of interferometry. Here's an artist's impression of the European project IRSI-Darwin.

9 Destination: earths!

In May 1998 the press announcement that followed the observation of a young binary in the Taurus constellation took the whole astronomical community aback. However, since it was based on observations by the Hubble Space Telescope (which was then the most sharp-sighted telescope available, able to resolve fine details of distant objects more than 10 billion years old) there was no reason to doubt it. Susan Tereby of the Extrasolar Research Corporation in Pasadena had aimed the Near Infrared Camera and Multi-Object Spectrometer at the Taurus constellation which is known to host numerous young stars and by chance observed a binary just at the moment of formation (it was a few thousand years old at most) at about 450 light-years from the Sun. And this binary exhibited some unexpected properties.

Streaming off this stellar couple is a long filament of luminous gas extending nearly 200 billion kilometres and ending at a bright, point-like object. Together they look like a sort of cosmic exclamation mark which perfectly illustrates the American team's puzzlement, though they searched for an explanation. It was suggested that the luminous point was a giant planet of 2–3 jovian masses, which, victim of the gravitational interactions between the two stars being born, could have simply been catapulted out of the system. The light trail could be matter from the protoplanetary disc trailed by the planet. If this were so, then Susan Tereby had obtained the first photograph of an exoplanet. Its surface temperature would be 1200 °C or colder. It was also possible, specified the press release, that the object was a little older, maybe 10 million years, and that in that case it would be a brown dwarf.

News of the discovery of the first visible exoplanet, which was named TMR-1C, spread rapidly round the world and was hailed by the mainstream media as a great triumph. However, astronomers were more sceptical; they felt the object needed further investigation. Which is exactly what Susan Terebey did by asking to be able to use the spectrograph on the Keck Telescope in Hawaii in order to learn more about the chemical composition of this mysterious object. The experiment was difficult to perform because since the object was very faint, it was difficult to focus on. Nevertheless, in 1999, it was shown that there is no water vapour in the object's atmosphere. For many, this showed that it was much hotter than previously thought, undoubtedly above 2200 °C, which is too hot for it to be a planet so far from its mother stars. Instead it must be a background star, placed by chance at the end of a gas jet which genuinely comes from the binary star. Susan Terebey accepted this conclusion in May 2000 after new spectroscopic observations.

In November 1999, another group claimed to have seen a planet. The head of the group was Andrew Cameron, a New Zealander from the University of Saint Andrews in Scotland. With the 4.2-metre William-Herschel Telescope installed in the Canaries, this team had followed the star Tau Boötes, located 50 light-years from the Sun and known to be accompanied by a burning-hot gas giant of 3 jovian masses, orbiting in 3.3 days. But how could the New Zealander's team have managed to see a planet so close to its central star? To be honest, they hadn't really seen the companion, but rather the light that it reflects from its star. This feat is possible in principle. To do so, you need to separate the few photons reflected by the planet from the flux of starlight. These can be recognised by using the spectral shifts induced by the rotation of the planet around its central star. This is not an easy task. The planet's spectrum is 20 000 times fainter than that of the star. This is an enormous contrast. You're at the limits of instrumental capabilities. Before Andrew Cameron, David Charbonneau of Caltech in Pasadena had tried to carry out this light extraction, and failed. Had the New Zealander succeeded where the Canadian had failed? It was thought so for some time, but during a conference in Manchester in August 2000, Andrew Cameron, who had continued his measurements, withdrew his initial conclusions. So, it seems that the companion of Tau Boötes hadn't been seen after all.

Other teams, like those formed by the Britons Matt Burleigh, Fraser Clarke and Simon Hodgkin, followed another path. Realising that the ratio between the luminosity of a sun and its planet, even in the near infrared, is overwhelmingly biased against the planet and makes it impossible to obtain a photo, they decided to get around the problem by looking for very faint stars.

They looked for stars at the ends of their lives, white dwarfs (mentioned in Chapter 5), nearly ten thousand times less luminous than the Sun. As white dwarfs started out as normal stars, the odds are that some of them have planets. There is just one problem: when a star reaches the end of its life, before turning into a white dwarf, it passes through the red giant stage during which it puffs up so much that it engulfs the bodies closest to it. In this way we expect that when the Sun reaches this stage, it will swallow the Earth and maybe Mars, but Jupiter will escape unharmed, even if it might be obliged to migrate to an orbit that is further out. At the end of the red giant stage, a star coughs a few times and throws out a lot of its matter into space. So, its gravitational hold on the bodies that go around it drops suddenly. In this way, the surviving planets of the red giant stage are either ejected into interstellar space, or else displaced to orbits further out.

This theoretical possibility for at least a few planets to survive the red giant stage and occupy outer orbits around a white dwarf was enough to satisfy our three British researchers, who early in 2002 started looking for gaseous giant planets of 3–5 jovian masses which should be only a few thousand times fainter than their dead stars. Their targets are the hundred white dwarfs within a radius of 50 light-years from the Sun. We are still waiting to find out if they are successful in their search for which they are using two of the most powerful existing instruments, the two Gemini telescopes.

But the planet hunters' favoured target remains main sequence stars, stars that are called 'normal', because it's around these that we

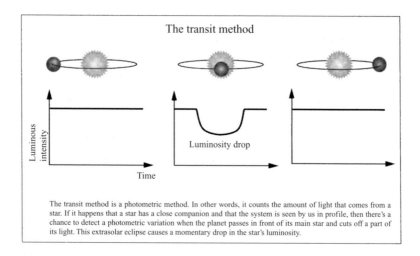

The transit method

Luminous intensity

Luminosity drop

Time

The transit method is a photometric method. In other words, it counts the amount of light that comes from a star. If it happens that a star has a close companion and that the system is seen by us in profile, then there's a chance to detect a photometric variation when the planet passes in front of its main star and cuts off a part of its light. This extrasolar eclipse causes a momentary drop in the star's luminosity.

can hope to discover planets that could harbour life. And this will undoubtedly require great ingenuity.

THE TRANSIT METHOD

Rather than seeing the light of a planet, we may, in 1999, have seen its shadow. The companion concerned is called HD 209458 b. It's 153 light-years from us, 'weighs' a half Jupiter mass and orbits its star in just 3.523 days. Three teams were able to watch the event, this sort of distant eclipse. First was our own, the Élodie group, and David Latham's group, with whom we shared our radial velocity data before giving everything to David Charbonneau and Timothy Brown of the National Center for Atmospheric Research in Boulder, who carried out the photometric measurements and discovered HD 209458 b's transit. That was during September 1999. A few weeks later, in November, the team composed of Geoffrey Marcy, Paul Butler and Steven Vogt sent their own radial velocity data to another expert in photometry, Gregory Henry, who saw the same phenomenon.

Each time we find an exoplanet stuck to its star, a hot jupiter, we check using photometry whether the star has periodical drops in its luminosity. If it does, the chances are that the system is right in our line of sight and that the drops in luminosity are due to the giant planet passing in front of the star. In 1995, we, of course, hurried to look for such an eclipse of 51 Peg by its companion, but we were unsuccessfully. Our system wasn't seen in profile. This was a pity, since that would have allowed us to be sure that we really had a planet and not just a star having a bout of tantrums or an instrumental problem.

There's no avoiding probability: there's a chance of about one in ten that an extrasolar system is in our line of sight and so about a chance in ten that we can see it in transit. Let's be clear that this only applies to exoplanets in the hot jupiter family whose orbital periods are less than 5 days. In fact, the closer a planet is to its main star, the higher are the chances of seeing an eclipse. All the gas giants with periods of several days discovered since 1995 have been subjected to such photometric searches. But these searches were in vain, until the discovery of the companion orbiting HD 209458, a star located in the Pegasus constellation.

Thanks to this transit and to additional data that it provided, such as the angle of the system with respect to our line of sight, we've been able to establish the precise properties of the planet. While it has a lower mass (by a factor of 0.69) than our Jupiter, it's one and a half times its size. This confirms the brilliant work by theorists like the Frenchman Tristan Guillot and the American Adam Burrows. They predicted that a newly formed gas giant, which is still puffed up by residual heat, doesn't diminish in size if it migrates quickly enough towards its star, where the ambient temperature brings the surface to 1500 °C, guaranteeing that it remains puffed up. The density of HD 209458's companion was also calculated to be 0.3 grams per cubic centimetre. If there were an ocean big enough to hold it, it would float.

After the announcement of this discovery, two astronomers from the Observatoire de Paris-Meudon, Noël Robichon and Frédéric Arenou, wondered if the astrometric satellite *Hipparcos*, which had observed many stars, including HD 209458, might have recorded something unusual about it. Indeed yes! Its detectors had recorded drops in luminosity five times, the first time being on 17 April 1991, but no-one had noticed.

Those who hunt planets breathed a great sigh of relief thanks to this discovery. It was the confirmation that there truly exist planets that are unbelievably close to their stars, and that these are gas giants and not giant solid telluric planets that would have had to have been constructed in place in a way that no-one could imagine. Our hot jupiters were not just the figments of our overripe imaginations.

But there's more to say about HD 209458 b, which keeps throwing up surprises. The Americans David Charbonneau and Timothy Brown announced at the end of November 2001 that they had obtained a glimpse of the atmosphere of this extrasolar planet. To do this, they used the Hubble Space Telescope and observed four transits of HD 209458 b. Each time the star is crossed, some of the light that it emits is absorbed by the chemical elements present in the planet's atmosphere, whose traces are accessible to the Space Telescope's spectrometer. The result is that the Americans detected the presence of sodium in the upper atmosphere of this distant world. The bad news, which was unsurprising given that the temperature of the planet is more than 1500 °C, is that the atmospheric envelope is unlikely to support life. On the other hand, the presence of sodium is good news for planetologists, even if a little less than was expected was detected, since it matches very nicely with the dominant theoretical models that describe these gas giants.

It goes without saying that when these planetary transits are detected, they are extremely rich in information. Another experiment, run by Ronald Gilliland for a week in 1999, looked for the transits of short orbital period planets in the globular cluster 47 Tucana using the wide-field camera of the Hubble Space Telescope. The idea was not to concentrate on one preselected system, but instead to observe a great number of stars in the cluster. In all, no less than 37 000 stars were checked. Preliminary studies had concluded that if the frequency of hot jupiters in 47 Tucana were comparable to that in the solar neighbourhood, then it should be possible to observe about 15–20 transits.

In June 2000, during a meeting of the American Astronomical Society, Gilliland and his team revealed their preliminary results for 27 000 of the 37 000 stars observed. Nothing had been discovered. There was no sign of a transit. Was this caused by an instrumental problem? The experts favoured another explanation. Not only are the stars in 47 Tucana poor in heavy metals, an annoying detail when you know that there's a link between high metallicity and planetary presence, but also the stellar concentration of the cluster must generate gravitational perturbations that are unfavourable to the birth and the survival of planets. We should add that this study had nevertheless provided highly valuable information on eclipsing binaries, variable stars and cataclysmic variables (couples formed from a normal star and a white dwarf, the latter emptying the former of its material).

At the Observatory of Geneva, we developed another method of observing transits, one that is spectroscopic rather than photometric. We start off with the curve defined by the spectral shifts of a star perturbed by a companion. Then we study the remainders of this oscillation. If the plane of the system lies in our line of sight, then the planet, when passing in front of the star, induces a sort of jump in the radial velocities of the star. Using this phenomenon, modelled by Anne Eggenberger, a young student at Geneva who hadn't even finished her degree when she carried out this work, we were able to detect the spectroscopic transit of HD 209458's companion.

It is also planned to launch satellites that specialise in observing photometric transits. COROT (an acronym of COnvection, ROtation and planetary Transits) is an essentially French project, coordinated by the Centre national d'étude spatiales (CNES, or National Centre for Space Studies). Instigated some fifteen years ago, it was initially destined to collect data in the domain of stellar seismology, the science which analyses pulsation modes of stars. This type of investigation requires a photometer able to measure stars for several months at high frequencies, and COROT's photometer was developed with this in mind. But with the discovery of exoplanets, researchers like the Frenchman Jean Schneider showed that COROT could work wonders in the detection of planetary transits.

Despite this, the project's prospects looked weak in 1999, when a political decision was made to relegate it to the list of low-priority missions. However, this decision has now been overturned and COROT should take off in around 2005. Once in space, it should carry out five observing campaigns of 150 days, each enabling data on about 10000 stars to be gathered. And while it is particularly intended to detect gas giants – it should even contribute to substantially increasing their statistical sample and thus contribute to better knowledge of their frequency – it is also hoped that COROT may detect telluric planets. To do so will require much care, since if we estimate that the transit of a jupiter in front of a sun would make its luminosity drop by about 1%, the effect of an earth would be a hundred times less.

Another project, that is similar but more powerful, comes from the European Space Agency (ESA). This mission, named Eddington, should be operational in 2007 at the earliest. Like COROT, this satellite will study both stellar seismology and planetary transits. No less than 500000 main sequence stars will be checked out over three years. And if everything goes as planned, Eddington should result in the detection of about a hundred telluric planets, and even better they should be located in habitable regions, where life is theoretically possible.

On the other side of the Atlantic, the USA has a project identical to COROT, except that it will be entirely devoted to the search for exoplanets. A satellite named Kepler, fathered by William Borucki, a transit pioneer, should from 2006 simultaneously follow 100000 stars in the solar neighbourhood, for four years. Its mission will be to explore the structure and the diversity of planetary systems. Of course, it is hoped that it too will detect telluric planets.

MICROLENSING

In 1936, the great Albert Einstein, who was rarely short of ideas, published an article in the review *Science* in which he made some comments on a phenomenon called gravitational lensing. This is deep in the domain of relativity since it involves masses that curve space-time and that, by doing this, also curve light rays. Imagine a distant star of

Microlensing

One of Einstein's great achievements was that he showed that the mass of an object is able to curve the structure of space. The result of this physical reality is that the path followed by light can be bent.

The microlensing effect is an example of this curvature. Light that reaches us from a very distant star is sort of diluted by its trip in the cosmos, and so appears very faint.

When an object passes between the distant star and the Earth, its mass forces the light from the distant star to be focussed towards the Earth. The star then seems, for a few hours, days or months, to be brighter.

Maximum brightness during a microlensing event

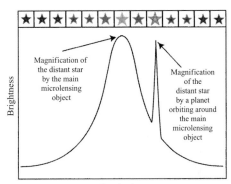

To discover a planet by this method, you have to point your telescope at a place in the sky densely populated by stars and try to catch microlensing events whenever you notice a star whose brightness changes.

You follow them day after day. If the object acting as the microlens is alone, you should observe just a single luminosity peak. If it's accompanied by a planet, there's a chance of seeing a second luminosity peak.

which you know the luminosity. By chance, while you're observing it, a massive body, another star for example, passes between you and the distant star. Locally, the rays are curved and concentrated in the direction of the Earth, where photometric detectors await them. The result is that the light of the distant star is amplified, sometimes by a factor of as much as 10, for a few days, or even a few weeks. You can even imagine a small black hole acting as a lens and having this effect for several months.

Even one planet alone can create the microlens effect, but it wouldn't be seen in time. However, if the planet is coupled with a star, the event would take longer and there would be a chance of discovering an anomaly in the light curve of the distant star that undergoes this magnifying glass effect. Like transits, microlensing events are able to reveal terrestrial mass planets. But while transits handle short periods better, microlenses favour planets further from their star, those at between 3 and 6 astronomical units.

An annoying detail is that any particular gravitational lens effect never occurs twice. The two objects involved in the phenomenon are generally quite far from one another, so they're not linked by any common force that could induce cyclic behaviour between them. It's mere chance that a massive object passes between a distant star and the Earth. It's up to astronomers to cope with the random nature of microlenses. Either they detect them in time, or else too bad! And this constraint is even stronger when the anomalies induced by a planet don't last for more than a day for a gas giant and barely more than an hour for a telluric planet of an Earth mass. All these difficulties explain why experts in the field have set up international observation networks linked to alarm systems. If one of the telescopes detects an interesting event, the information is passed by email to observers in other countries, who hurriedly point their instruments to the coordinates indicated. In this way, if a cloud gets in the way of one telescope or even two, there's a fair chance that the rest will be able to continue to watch under clear skies.

This technique wasn't invented just for finding exoplanets. The first programmes to have used it – that of the Australo-American team MACHO (MAssive Compact Halo Objects), of the American–Polish team OGLE (Optical Gravitational Lensing Experiment) and of the French team EROS (Expérience pour la Recherche d'Objets Sombres, Experiment for the Search for Dark Objects) – were put in place in order to look for dark matter in the Milky Way, the missing matter without which you can't explain the rotation speed of our Galaxy. It was Bohdan Paczynski, a professor at the University of Princeton and director of the OGLE programme, who, in 1986, mentioned the possibility that the missing matter could be composed largely of compact dark objects that are massive enough to induce many, short, microlensing events. You just had to point a telescope towards the central bulge of the Galaxy or towards the Large Magellanic Cloud, where the concentration of stars is very high, in order to increase the chances of seeing the phenomenon. The Galactic Centre is best seen from the southern hemisphere, which explains why the telescopes involved in these projects are in South Africa, Australia, New Zealand and Chile.

Since the beginnings of these programmes, hundreds of microlensing events have been detected. But none of them bore unambiguous signs of a planet. It was long thought that the key event was that named MACHO-97-BLG-41, that is, the forty-first microlensing event detected in the MACHO Galactic Bulge programme during 1997. The discovery was announced in November 1999 by the Microlensing Planet Search (MPS) team, led by David Bennett and Sun Hong Rhie of the American University of Notre Dame. According to MPS, in the case of MACHO-97-BLG-41, which occurred 20 000 light-years from the Sun, the model which seemed to work the best was a scenario consisting of a stellar couple with a planet orbiting around it. The only cloud in the picture was that MPS's competitor, the PLANET group, had also observed MACHO-97-BLG-41 on 23 July 1997. Their conclusions were different: they said there was no need to introduce a planetary hypothesis since the behaviour of a binary alone was enough to explain the anomalies in the light curve. It seemed the small suspicious peak was just a consequence of the two stars rotating around one another.

A disagreement on an experiment that is impossible to reproduce is something that is certain to lead to a certain scepticism. The more so when a second event, named MACHO-98-BLG-35, announced in January 1999, also led to disagreement. For the MPS group, the anomaly detected in the light curve was due to the presence of an exoplanet slightly less massive than the Earth and located at 2 astronomical units from its star. The PLANET team was more cautious. It didn't contest the existence of an anomaly in the curve, but said the anomaly was so weak that it was at the threshold of detectability. It wasn't possible to tell if it was an instrumental problem or a blip from the terrestrial atmosphere.

More recently, it was decided to add another arrow to the bow of the OGLE programme, which is normally oriented towards microlensing events, by using it to try to find planetary transits (see 'The transit method' above) in front of distant stars. After all, the two methods are rather similar. They both rely on changes in luminosity to capture their prey and do this by using photometry. However, while one looks for increases in luminosity, the other looks for drops in luminosity.

By using this second method, the Pole Andrzej Udalski and his colleagues from the University of Warsaw announced, in February 2002, the discovery of 46 transits after having scrutinised thousands of stars close to the centre of the Galaxy. These observations were carried out with an instrument installed at the Las Campanas Observatory, in Chile, and then reanalysed using a new, more efficient method. Not only did the new method find the 46 transits we just mentioned, but it also led to the discovery of 13 new ones in June 2002, 13 planetary candidates of which one, OGLE-TR-56, seems to have a mass about that of Saturn.

As promising as they are, these discoveries still suffer from a slight uncertainty. Photometry, which was used to track these transits, measures the variation in luminosity and deduces the size of the object that induces the eclipse. However, it's possible that the transit is induced by an object larger than a planet, but that this object only eclipses a part of the main star. One way to remove this ambiguity – and we do this with the Very Large Telescope – consists of measuring the star's radial velocity in order to obtain a good estimate of the companion's mass. In science, you often have to have many different methods available in order to be sure of a result.

Despite this, we're still not satisfied. Low mass exoplanets are out of range for the radial velocity method. They were sought by those who like the microlensing method, but they've yet to be found.

Programmes like MPS, OGLE and PLANET still have a lot of work ahead of them. And they're not alone.

TERRESTRIAL INTERFEROMETERS

In the race to detect exoplanets, the radial velocity method outstripped the astrometric method. At the time, the latter lacked the power to achieve its goals. It was always pushing the limits of its detection ability. That epoch is over. Astrometry once again has a rosy future. It's preparing its revenge and is about to overthrow the radial velocity technique in order to see more and further. It can today count on the support of the two biggest terrestrial telescopes working in the visible wavelength domain: the Keck, installed on the peaks (at an altitude of 4200 metres) of the island of Hawaii, with its two giant 10-metre mirrors, and the Very Large Telescope (VLT), the property of the European Sourthern Observatory, which is installed with its four 8.2-metre mirrors at Paranal, Chile, at an altitude of 2650 metres and offers one of the purest skies on the planet. One of the characteristics of these two giants of terrestrial astronomy is that they were conceived to function in interferometric mode, a process which is particularly useful for planet hunters, and whose extraordinary potential was shown by the first tests made in 2002 on these two machines.

Interferometry is based on the wave nature of light: light crosses space like a wave. If two waves meet, they interact and can be thought of as a resulting third wave (we also say that they create interference). The characteristics of the resulting wave depend on both the properties of the two original waves and the way in which they met each other. If they have exactly the same wavelength, then there's a chance that their respective peaks and troughs are exactly in phase. We call this constructive interference: since the resulting wave has a greater amplitude, it's more luminous in some sense. In contrast, if the peaks of one wave line up with the troughs of the other, then they cancel each other. We call this destructive interference. This is how two luminous waves, by interfering in a certain way, can result in total blackness.

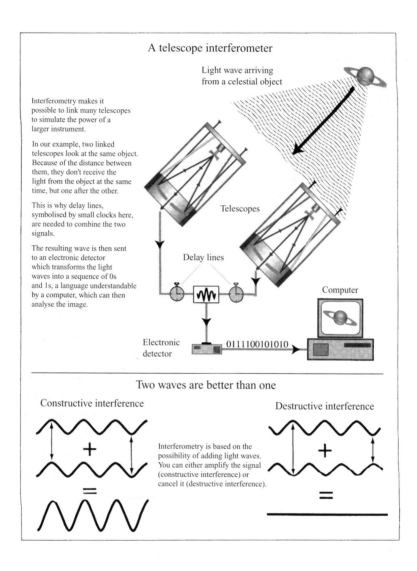

A telescope interferometer

Light wave arriving
from a celestial object

Interferometry makes it
possible to link many telescopes
to simulate the power of a
larger instrument.

In our example, two linked
telescopes look at the same object.
Because of the distance between
them, they don't receive the
light from the object at the same
time, but one after the other.

This is why delay lines,
symbolised by small clocks here,
are needed to combine the two
signals.

The resulting wave is then sent
to an electronic detector
which transforms the light
waves into a sequence of 0s
and 1s, a language understandable
by a computer, which can then
analyse the image.

Telescopes

Delay lines

Computer

Electronic
detector

0111100101010

Two waves are better than one

Constructive interference

Destructive interference

Interferometry is based on the
possibility of adding light waves.
You can either amplify the signal
(constructive interference) or
cancel it (destructive interference).

This property of light is a help to astronomers, who are always greedy to see more and further. A telescope is characterised by its surface area and its diameter. The greater the collecting surface, the easier it is to detect very faint objects. The greater the diameter, the better you can make out the details of an object. To profit from these to advantages, it's enough to make huge mirrors. But the problem is that the cost increases exponentially with size. Lack of unlimited financial resources means that researchers have to rely on their ingenuity to get around the problem and rather than increasing both area and diameter they've increased just the latter. Taking two small telescopes and separating them by 10 metres gives the same angular resolution as a 10-metre mirror, as long as the two light waves are correctly combined.

This technique, invented by the American Albert Michelson (1852–1931), and for which he received the Nobel Prize in physics in 1907, first met with great success in the domain of radio astronomy, where the wavelengths are relatively large (you can count them in centimetres), which makes it easier to get the technical apparatus for putting the waves into phase to work. As a rule of thumb, the precision of an interferometric set up should be equal to a tenth of the wavelength that it uses, so you can now see how difficult it is to carry out optical inteferometry in visible wavelengths which are counted in micrometres, that is, in millionths of a metre.

However, this is what the interferometer at the VLT, the giant European telescope in Chile, will do. In time (probably before 2010), and if the money is available, it should be possible to combine the four 8.2-metre telescopes and the three small 1.8-metre auxiliary mobile telescopes together. But planet hunters won't need all of this power. Two telescopes functioning as an interferometer will be enough for them to carry out high precision astrometry. Most of the work will be done in narrow angles, in extremely tiny portions of the sky. Telescopes will be pointed at a group of just a few stars. Typically, three will be enough, the first two serving as references, the third being the target. You can see the value of interferometry for this. Thanks to the increased angular resolution, it's possible to detect much smaller changes in stars' positions than before. While classical astrometry flirted with a resolution of 10 milli-arc seconds, the VLTI, or in other words, the interferometer version of the VLT, aims at reaching 10 micro-arc seconds between two objects. That's a thousand times better. This should enable the detection of star–jupiter systems up to distances of the order of 1 kiloparsec (i.e. 3260 light-years) and lighter planets, of ten or so terrestrial masses, around the closest stars to the Sun.

With the Keck interferometer, the Americans hope to attain results in the same ball park. Like the Europeans, they have to cope with the stringent requirements of metrology. Focussing telescopes separated by dozens of metres and doing so to a precision of a few nanometres is one of the toughest technological challenges.

So it's in the domain of astrometry, an indirect detection technique, that Keck and the VLT will carry out most of their planetary harvest. This won't stop them from also exploiting their interferometric power to try to directly see exoplanets around close stars. It goes without saying that these searches will be carried out in the infrared, where the luminous balance between stars and planets is the least unfavourable to the latter. However, even for Keck and the VLT, the game is far from being won in advance. They'll be forced to work at the limits of their abilities. Their targets, hot and luminous, will be gas giants close to their star and young planets, a task requiring exposure times of several hours.

While the future for terrestrial observation lies in interferometry, this doesn't mean there will be no further use for classical telescopes with a single antenna. Two small telescopes can together equal the angular resolution of a 100-metre diameter telescope, but they can't attain the sensitivity of a mirror with the corresponding surface area. It's undoubtedly for this reason that the European Southern Observatory has conceived a huge contraption for the future, named OWL (OverWhelmingly Large telescope). This huge 100-metre bird of prey is inspired by the mirrors of the Keck, which, in contrast to those of the VLT, which are each made as a single piece and supported by hundreds of hydraulic lifts, are instead made of dozens of small, perfectly adjusted and coordinated hexagonal mirrors. This approach, though technically arduous, has now proved its efficiency and makes it possible to consider terrestrial mirrors several dozen metres in diameter (even if these are very expensive). OWL should be able to see details as small as a milli-arc second, but without having to bother about the long exposures that characterise interferometers, though for the moment it only exists on the drawing board.

SPACE-BASED INTERFEROMETERS

As you can see, terrestrial interferometry is no picnic. But the experts, either unaware of the problems or because they like challenges, have decided to go further, higher, and to develop spatial interferometry. They'll have to find not only a method of sending the interferometer into space without it being totally put out of shape by the constraints of a rocket trip, but also a way of synchronising the mirrors with the same precision required on Earth.

Despite the difficulties, the Americans are planning such an experiment starting in 2008 with the launch of SIM (Space Interferometry Mission) that is planned to last from five to ten years. This satellite will take the form of a small, initially foldable, but finally rigid, beam, 10 metres long with a small, roughly 30-centimetre mirror at each end. Despite their small size, the mirrors should accomplish marvels by benefiting from the purity of empty space and the power of interferometry. They'll make it possible to attain an astrometric precision of 1 micro-arc second, where the great terrestrial interferometers will at best only achieve 10 micro-arc seconds. As its proponents point out, SIM could observe grass growing second by second in a field 10 kilometres away.

Planetary specialists expect a lot from such an instrument, which, we should point out, will also work in many other domains. SIM will be able to detect jupiters orbiting around stars as far as 3260 light-years and could thus find a few hundred. For lighter planets, between 5 and 10 terrestrial masses, it will concentrate on 200 of the stars closest to the Sun. It could detect an extrasolar earth provided that the main star is no further than 10 light-years from the Sun and that it's in the M dwarf class, that is, it has a low mass.

The experience acquired from SIM will be very useful for the succeeding missions by NASA in its programme *Origins*, which aims to find extraterrestrial life before the second half of the twenty-first century. SIM, by obtaining a rich harvest of planets, will enable its successors to better choose their targets. These successors will not only be more powerful, but they'll also use different flight plans. As we saw, SIM will probably be based on a rigid, deployable structure that will play an important role in the precision and the synchronisation of the mirrors. While this is possible with a small satellite, it becomes unthinkable with an instrument that includes several mirrors separated by larger distances. So astronomers and engineers have thought of another solution. Why not create a space interferometer the components of which are not physically attached to one another, but instead float freely in space?

As incredible as this may seem, this is not impossible. To prove this, American experts from NASA are working on an experiment planned for 2005. Named StarLight, this satellite will be composed of two small, independent telescopes freely floating in space, up to one kilometre from one another. Despite this great separation, the two mirrors will have to be synchronised to within a few centimetres, a feat that will be partly guaranteed by a metrology system, based on lasers that will continuously control the separation, commanding the mirrors to reposition themselves whenever necessary.

This metrological requirement might seem exaggerated, but on the contrary, it's unavoidable for the pursuit of an ambitious programme, such as that of the great space interferometers. Today there are two projects. One is American and has the eloquent name of Terrestrial Planet Finder (TPF). The other is European. Thought up by the Frenchmen Alain Léger and Jean-Marie Mariotti, it was first called Darwin and then renamed the InfraRed Space Interferometer (IRSI). The specifications have not yet been set, but we can guess that the two floating interferometers will gather together five or six independent mirrors (of about 1.5 metres in diameter) separated by a few dozen metres. In contrast, what is sure, is that TPF and IRSI will work in the infrared, an indispensable condition for the success of their primary mission, which is to see terrestrial planets around the closest stars to the Sun and to use spectroscopy in order to learn more about the composition of their atmospheres. And who knows if one of them will show indirect signs of biological activity!

These two projects, if they are accepted, won't be operational before 2020. The thorny question of funding them will obviously have to be dealt with, since such a project will be costly, to say the least. Undoubtedly the two space agencies have understood this, so after preliminary negotiations have not ruled out the idea of collaborating and uniting their forces to guarantee the long term success of such an interferometer. It wouldn't hurt, of course, to have two teams taking up a challenge of this sort. The goal is worth it. How fantastic it would be to admire the photograph of an extrasolar earth, even if, on the photo, the planet was just a tiny, brilliant point.

It remains to be seen how TPF and IRSI plan to achieve this feat. Once again, they'll use the infrared, which makes it possible to reduce the difference in luminosity between the star and the planets. However, this wouldn't be enough to achieve the goals – it will be necessary to use another technique. Usually called nulling and discovered by Ronald Bracewell of Stanford University, this technique has already shown its colours on various instruments such as the VLT and Keck interferometers and on the SIM satellite, and will hopefully show its true glory in the great space projects.

Nulling is an ingenious and efficient way of enabling a planet to be seen despite the outsized luminous ego of its star. Like interferometry, of which it's a special case, it's a practical result of the wave nature of light. Imagine two telescopes, functioning interferometrically, aimed at the same star. Each receives a part of the stellar light. In interferometry, you're mostly interested in getting the two light beams in phase in order to create positive interference. Nulling uses the opposite effect. Before combining the two beams, an optical trick inverts one of the beams, which becomes something like the negative of the other. The peaks and troughs of one face the troughs and peaks of the other, respectively. As a result, when they're added together, the waves cancel out, neutralising each other. Blackness results.

The more mirrors there are in the interferometer, the better this mechanism functions. As will be the case for the TPF and the IRSI, these instruments should be able to extinguish enough light from the stars to allow that of small planets like the Earth to appear. This ability to extinguish a star without at the same time masking its planets comes from the fact that only the wave front that comes from the target object, in this case the star, is removed. Even if it's tiny, when seen from the Earth, the distance between a main star and its companion is enough for the light from the latter to arrive at the mirrors at a tiny angle which stops it from falling into the nulling trap. To sum up, you obtain, on the one hand, destructive interference which extinguishes the star and, on the other, constructive interference which makes it possible to see the companion.

Thanks to this technique, it should be possible to see terrestrial planets orbiting around their stars. The conditional tense, 'should', is critical here because an unknown factor remains. Even if nulling works, even if the coordinated flight succeeds, even if metrology attains the required nanometric precision, it is still possible that TPF and IRSI will see nothing other than a reddish halo which is induced by exozodiacal light and makes everything opaque. This strange light in our own Solar System can sometimes be seen from the Earth, for example, during a sunset seen from the mountains. Just a few instants after the Sun disappears, a luminous band appears in the already darkening sky. This is due to interplanetary dust spread through the equatorial plane of the Sun, which, for a brief moment, plays with its rays. In the domain of visible wavelengths, this zodiacal light is particularly flimsy. In contrast, in the infrared it becomes an astronomer's worst enemy. These tiny grains heated by the Sun blind any telescope that tries to observe a large portion of the sky in the infrared, whether space-based or not, interferometric or not. This is why it's planned to send IRSI very far from the Earth, to a distance of about 4 or 5 astronomical units, in Jupiter's back yard, where the amount of dust is negligible.

However, such a manoeuvre, while avoiding the zodiacal light from our Solar System, can't do anything about the exozodiacal light, in other words, about a similar halo of dust that could surround a distant planetary system which would prevent the detection of planets in the infrared. We still don't know today if such light exists around each planetary system or if it's as intense as in our own. We hope to answer this question, thanks in particular to the SIM satellite, which will study protoplanetary discs in detail.

SURVEYING LIVING ATMOSPHERES

Let's be optimistic and imagine that the exozodiacal light is not intense enough to be a problem or that we find a technical trick to get around it. We would then be able to see the silhouettes of extrasolar earths. Once again, they would be just small luminous points, but these simple images would constitute one of the most extraordinary episodes in the history of astronomy. It would be a real scientific achievement, but the experts would not rest on their laurels. On TPF as on IRSI, they'll have taken care to install a powerful spectrograph whose role will be to analyse the chemical signatures of the planets' surfaces. Using this, we'll be able to learn whether or not a planet has an atmosphere and whether or not it harbours life.

The aim may appear ambitious, and it is. The main difficulty in the project is time. Patience will be needed to carry out such an analysis. For days, maybe even weeks, the telescope will remain focussed on the same star in order to gather enough planetary light for its spectrograph. But once the project is finished, the information obtained will be especially valuable.

As reference models, astronomers use what they know about the different planets of the Solar System and of one of them in particular, the Earth. After all, and at least for the moment, it's the only one that we know to host life. With just one example, it's difficult for us to imagine life that could be radically different from ours. And when we say different, that doesn't just mean that we might find dragons with seven legs on another planet, but rather that these same dragons might be 'constructed' differently, they might not share the same carbon chemistry that governs all forms of life on Earth, from the simplest to the most complex.

Therefore, abandoning the idea of a world completely different from our own, it's better to start from the principle according to which other life needs the same conditions as those on Earth, conditions which we can read off from our own atmosphere. This was the experiment carried out by the American Carl Sagan (1934–1996) of Cornell University.

Carl Sagan is, of course, famous for his magnificent popularisation of science: his books have attracted many into astronomical careers and inspired them with the desire to discover life elsewhere in the Universe. He worked on space missions as famous as *Mariner 9* and the two Vikings which went to the planet Mars in the 1960s and 1970s. It was also his idea to put the famous plate, showing a man and the position of the Earth with respect to the Solar System and the Galaxy, on the Pioneer 10 probe, in case an intelligent extraterrestrial – hopefully friendly – comes across this testimony of terran technology.

Carl Sagan was a great planet specialist. He made it possible to better assess Venus and its terrible greenhouse effect maintained by a thick carbon dioxide atmosphere, which sustains an average ground temperature of over 450 °C. He also predicted that the atmosphere of Titan, Saturn's largest moon, had to contain organic molecules similar to those that participated in the buildup of life on Earth, and this was before the Voyage 1 and 2 probes confirmed his hypothesis in the 1980s. One of his final contributions was to propose that the American probe Galileo, launched in 1990 with Jupiter as its destination, should take advantage of its acceleration as it passed the Earth to carry out a spectroscopic survey in the near infrared. This survey could serve as a working model for those who hope one day to analyse the atmospheres of exoplanets.

Galileo's data revealed three properties that characterise life on Earth. First of all, there's an absorption line at a wavelength of

0.7 micrometres which outlines the continents. It's a line due to chlorophyll, the green coloured pigment present in plants that plays a dominant role in photosynthesis. Another, particularly sharp, absorption band, is clear at 0.76 micrometres: this is that of oxygen gas (O_2), which makes up 21% of our atmosphere. This is a much higher proportion than has been detected on other planets of the Solar System. While certain abiotic processes can produce oxygen, for example, the dissociation of water molecules (H_2O) by the Sun's ultraviolet rays, they are nonetheless unable to produce the quantities observed in our atmosphere. The only satisfactory explanation is respiration by plants, which consume carbon dioxide and breathe out oxygen.

Finally, the third property recorded by Galileo is the signature of methane (CH_4), in the proportion of one part in a million. But why make a fuss about so little when we know that objects like Jupiter and Titan contain considerable quantities of methane? It's because on Earth, this gas is unexpected. In the presence of atmospheric oxygen, it reacts with the oxygen to form water molecules (H_2O) and carbon dioxide (CO_2). Therefore, if methane is detected in terrestrial air, there must be a continual process that produces it. This source is methane-producing bacteria which populate swamps, rice paddies and cows' guts.

In short, it would be enough for the great space interferometers to detect these three absorption lines in an exoplanet's spectrum to conclude that life exists in its lands and oceans. Unfortunately, several pointers suggest that such detections would be ambiguous. You have to watch out for traps. The presence of oxygen alone would not guarantee a living source. As we saw above, the atmosphere of a planet can contain small quantities of oxygen due to photodissociation of water molecules by the ultraviolet rays from its star. Because of this, it's possible that during its youth, Venus had a substantial quantity of abiotically generated oxygen. This phenomenon only occurs on a large scale in very wet stratospheric layers. But those of the Earth are mostly dry, so the spectral signature of atmospheric water should help to clarify the origins of the oxygen.

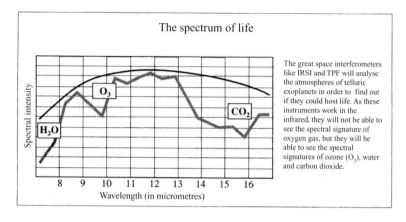

The spectrum of life

Spectral intensity

O_3

CO_2

H_2O

8 9 10 11 12 13 14 15 16
Wavelength (in micrometres)

The great space interferometers like IRSI and TPF will analyse the atmospheres of telluric exoplanets in order to find out if they could host life. As these instruments work in the infrared, they will not be able to see the spectral signature of oxygen gas, but they will be able to see the spectral signatures of ozone (O_3), water and carbon dioxide.

A more annoying detail is that both IRSI and TPF will be blind to oxygen gas, the famous O_2 so characteristic of life on Earth. Its absorption line, though seen by *Galileo*, will be invisible for these future space interferometers which will use other, infrared, wavelengths. Luckily, there's an easily accessible substitute: ozone. Rather than being made of two oxygen atoms, as is the case for oxygen gas, ozone (O_3) has three. It has an absorption line at 9.6 micrometres, a wavelength within reach of IRSI and TPF. So we can breathe a sigh of relief, oxygen is again detectable, even if indirectly, thanks to its cousin. The only drawback is that ozone is not a linear indicator of oxygen. In other words, a relatively strong line at 9.6 micrometres does not necessarily correspond to a large amount of oxygen gas. Here again, the detection has to be combined with others (that of atmospheric water, for example), in order to be sure that the ozone is not the by-product of abiotic oxygen.

Just as the presence of oxygen is a strong, but ambiguous, indicator of life the opposite is also true. The absence of oxygen doesn't guarantee that a planet is lifeless. According to geologists, oxygen reached significant levels on the Earth about 2 billion years ago. Yet according to the fossil record, life appeared no later than 3.5 billion years ago. This means that living things inhabited the Earth for nearly 1.5 billion years before the oxygen level was detectable in the atmosphere.

The first single-celled organisms didn't need oxygen to live. They took their energy from carbon dioxide and from molecular hydrogen in which the Earth's primordial atmosphere was relatively rich. In exchange, they ejected methane which was able to build up without much trouble since there was no oxygen to break it up into water and carbon dioxide, as we described earlier. A strong methane signature would constitute a clue to the presence of methane-producing organisms on a planet. This is a signature that TPF and IRSI would not fail to detect. With a wavelength of 7.6 micrometres, they should see it clearly.

The first criterion in deciding whether a planet has the potential to host life is to find out if it lies in the habitable zone, in other words, if it's located at the right distance from its star to allow the existence of liquid water, an indispensable condition for the appearance of life. It's primarily this sort of planet that the great space interferometers will look for, relying on their predecessors to show the way.

GAIA, a key project of the European Space Agency, will not have the power of IRSI. But it will go beyond the *Hipparcos* satellite, which, between 1989 and 1993, measured the celestial coordinates, the annual parallax and the proper motion of 120 000 stars with a precision ten to a hundred times better than previous measurements. This satellite, whose funding was agreed in October 2000, should be launched within a decade. Once in orbit, it will measure several hundreds of millions of stars with a precision of micro-arc seconds.

This will definitely teach us a lot about the origin, the structure and the evolution of the Milky Way. But we also expect GAIA to yield a rich harvest of thousands of exoplanets and brown dwarfs. This work will make it possible to determine the proportion of stars of our Galaxy that shelter planets. And while GAIA won't be able to detect terrestrial mass planets, it will sort out extrasolar systems into those with gas giants whose positions allow the possible existence of telluric planets and those that don't. Undoubtedly the space interferometers will be particularly indebted to this groundwork.

In the space domain as well as in ground observations, the arrival of interferometry won't relegate classical single mirror astronomy to the dustbin. Proof of this comes in the worthy successor to Hubble, called the Next Generation Space Telescope (NGST), a joint project of Europe and the USA whose launch is planned for 2008. It will be much bigger than its older sibling. There is talk of a light cooled mirror 8 metres in diameter, compared to 2.4 metres for the Hubble. Also, in contrast to the Hubble, the NGST will not use visible light but infrared instead. Its images will undoubtedly be less spectacular than the magnificent photographs from Hubble, but its fantastic power will open hitherto unexplored regions of the Universe. It goes without saying that, in these wavelengths, it should be a formidable hunter of cold objects like brown dwarfs, protoplanetary discs, and, who knows, with a bit of luck, some young and massive jovian planets.

TAKING A PHOTO OF ANOTHER EARTH

What a project! With SIM, NGST, IRSI and TPF, exoplanet astronomy is planning a very promising future for itself right up to 2020. Yet, astronomers are already planning further ahead, because they have to allow for the time needed to envisage, design, finance, build and use a telescope. Looked at in this way, 2020 is tomorrow. So, though IRSI and TPF have not yet been agreed, astronomers have already started imagining their successors that would be launched around 2025–2030.

But what more could you want to see? Can you do any better than detect planets? Yes, is the astronomers' answer. You can try to see the planets, really see them.

An American outline exists for a project called the Planet Imager. It's an interferometer composed of five elements, each carrying four 8-metre diameter telescopes. It's like sending five Very Large Telescopes into orbit and then spreading them out in order to imitate an antenna 6000 kilometres in size.

There is an even more ambitious European project: that of the Frenchman Antoine Labeyrie, whose fertile mind is never short of new ideas and inventions. For him, a hypertelescope able to provide a resolved image of a small exoplanet isn't just science fiction, provided, of course, that the necessary technical and financial resources are available. Antoine Labeyrie's idea for this uses a technique he invented, diluted optics, which makes it possible to network hundreds, even thousands of mirrors. The coordinated flight of this mirror network would be powered by solar sails. Once fixed on a target, this telescopic armada would make it possible to obtain a resolved image of a distant telluric planet.

However, to achieve such a performance, it isn't enough just to use diluted optics. An *a priori* unsolvable problem had to be dealt with first. In order to obtain an instantaneous image using a great interferometer like that envisaged by Labeyrie, there would need to be a huge number of mirrors. And this requirement leads to a drawback, that of creating a lot of diffraction, which fuzzes up the image. Imagine, for example, as the French expert reminds us, how difficult it is to photograph a countryside through a curtain with a fine mesh. As the light goes through this mesh, it diverges, creating a luminous veil which hides the details of scene. And the smaller the size of the holes in the mesh relative to their spacing, the stronger this annoying phenomenon is. Given this rule, it was difficult to imagine much future for these great spatial interferometers that use diluted optics, Until Antoine Labeyrie came up with a solution. Using an ingenious combination of lenses, he succeeded in making the beam of light sent by each of the mirrors towards the central collecting station denser, so that the final image is much more luminous. The simulations and the calculations leave no doubt, the gain can attain a factor of a billion. There is just one drawback: the field size of the image is reduced, a fault which can be rectified by adding more mirrors.

Antoine Labeyrie's work has already convinced many and has influenced the creation of instruments like TPF and IRSI. But the Frenchman is already thinking of a more grandiose spatial interferometer, a hypertelescope 150 kilometres in size, named the Exo-Earth Imager, and comprising 150 mirrors of 3 metres in diameter, capable of providing a resolved image of a planet like the Earth located at a distance of 10 light-years. We could then see the colours, the contours of the continents, the clouds, and maybe, who knows, the signs of extraterrestrial civilisations, if they happen to fancy works of art as ambitious as the Great Wall of China.

That is the Holy Grail, the hope which tempts all planet hunters: to find life elsewhere, to show that it's not unique to the Earth, that it's spread more or less all over the Galaxy and across the whole Universe.

10 Further yet: life

Will we find extraterrestrial life? Has another planet in the Universe succeeded in assembling the extraordinary rainbow of conditions that life seems to require in order to appear? This is the ultimate question that lies behind the quest for exoplanets. But even if you're a member of the club of those who believe that life is not a terrestrial privilege and that it has undoubtedly developed on other planets, in other solar systems, there's still a challenging question: *how* will we find extraterrestrial life?

If one day humanity succeeded in finding such proof, we would confront the fourth cultural shock of our history. After having learnt from Copernicus that we're not at the centre of the Universe, from Darwin that we're the 'descendants' of an ape who herself is the very distant grandchild of a simple cell, and from Freud that we're subject to the whims of our subconscious, we would also have to cope with the idea that we're not the only living beings in the Universe.

Based on our present knowledge, it's becoming more and more difficult to imagine that the Earth is the only host of life in the Cosmos. In our Galaxy alone, there are more than 100 billion stars, while there are billions of galaxies in the observable Universe. Why would life have contented itself with appearing on a single planet, as beautiful and blue as it is? Each star is born in the same way, by the fragmentation of an interstellar cloud, and creates around itself an accretion disc. Even if only a small fraction of these discs give birth to planets, this would mean that there are billions of planets and it is easy to imagine that some of them, helped by chance, have got together the necessary ingredients for the appearance of life.

Obviously, we still have to find this extraterrestrial life. The easiest way would be to detect radio signals from another civilisation.

Their artificial character would be easily identifiable and would leave little doubt as to their origin. But up to now, the (mostly American) programmes for Searching for ExtraTerrestrial Intelligence (SETI) have been unsuccessful. There are so many possible stars, so many possible wavelengths to search! It's a colossal task, and those who've taken it up rely on planet researchers to help them to target their searches. In particular, they would like to know around which stars telluric planets have been detected in the habitable zone, i.e. the zone around each star, depending on its luminosity, where water can exist in the liquid state.

WHAT IS LIFE?

But what would extraterrestrial life be like? Like single-celled creatures similar to those which inhabited the Earth during its first two billion years? Like more evolved, maybe even intelligent, organisms? In this domain, science fiction gives itself free rein, but science has less freedom to play with. It has to conform to the laws of the Universe. Just think of terrestrial attraction and atmospheric pressure, terrestrial life has to take these into account. Muscular mass, skeletal structure, all these mechanical options are not merely fortuitous. Birds have hollow bones in order to help them fly and optimise their energy budgets. Evolution and adaptation yield to the requirements dictated by the environment. And there's no reason for this not to be the case in any other corner of the Universe.

In short, life, even if it does have a huge field of possibilities, cannot spring forth wherever it feels like it, developing without the merest constraint. This is the ransom paid for its extraordinary complexity. To describe a single cell, even the most basic of all bacteria, requires infinitely more information than is needed for any inanimate object. Why such complexity? Undoubtedly it is due to the functions that characterise life. It has to grow and multiply. This requires it to maintain itself in a relatively stable and permanent state. The price to pay for this is a significant spending of energy. And consumption of energy requires fuel. As the fuel is rarely directly available in the preferred form – just think of crude oil, which has to be refined and transformed into order to make petrol for your car – it's up to the cell to play the role of a refinery.

A cell doesn't just have to guarantee its own survival. It also has to contribute to the survival of its species. Because evolution didn't feel it was right to favour immortality, individuals come along and then pass away, for the good of the species, which succeeds using the succession of generations and successive mutations to adapt to an especially unstable environment. To die in order that others might live, in order that the species perpetuates, is the equation resolved by reproduction and replication.

This is a key function of life, one of those that lets you distinguish animate from inanimate, even if the distinction is not always obvious. What can you say about crystals, for example, members of the mineral world, but which grow and reproduce their structures? Are they alive? No, the structure is not complex enough, the pattern reproduced is unique and it can't adapt to the environment. A cell, on the other hand, can adapt. Reproduction, evolution, homeostasis in other words, the ability to maintain a structural and functional equilibrium, these are the pillars of life. There are nonetheless some difficult questions. What is the status of viruses? They have membranes, they evolve, they contain the necessary information for their replication, though they're incapable of carrying out the latter without the help of the cells which they feed off. Without their hosts, they can't exist. So, are they living or not? It is a question of judgment. Once again, the frontiers are fuzzy.

It's clear that life is complex. In its functions as much as in its structures. An incredible number of chemical actions are needed to construct a single-celled organism. The carbon atom is the architect of this complexity. It's at the centre of everything. It's the skeleton of life. This is because its intrinsic qualities make it a particularly sociable atom. Thanks to the four electrons occupying its second orbital, a carbon atom offers four possible bonds. Not only does it get along well with other elements, but it also gets on perfectly with members of its own family. So it can create long carbon chains to which other essential atoms for life like hydrogen, oxygen and nitrogen can attach themselves. This is what the Lego of life owes its complexity to: to multiple chemical elements, to a great variety of possible marriages and to an astronomical number of possible matches. All of this is thanks to carbon.

Could another element satisfy this requirement of complexity, diversity and wealth? Silicon, the only other element in the periodic table that, like carbon, has four free electrons, has sometimes been mentioned. And that's not the only argument that suggests it's a possibility. On the Earth, it's also nearly 135 times more abundant than carbon. Did life bet on the wrong horse? Probably not. Because despite its good looks, silicon has a ghastly fault. Even if gets on well with others, it doesn't have a well-developed family tradition. The bond between two silicon atoms is only half the strength of that between two carbon atoms, which discourages the formation of the long molecular chains typical of life.

Carbon is very probably an unavoidable ally for life. As it exists in many places in the Universe, we can ask whether life is a frequent phenomenon. To answer to this question we need to examine the conditions that permitted the hatching of life on Earth.

THE EMERGENCE OF LIFE ON EARTH

Our blue planet has not always been like it is today. About 4.55 billion years ago, when it was newly formed, it was just a big incandescent ball. Then, for a few hundred million years, it had to cope with an unending meteoritic downpour. Hot and bubbling, our Earth was nothing but lava and magma, volcanoes and the smell of rotten eggs, gas bursts and vaporisations, before it cooled, when rain started to fall and create bodies of water. Somewhere, in some remote corner of a primordial ocean, a few molecules, by chance, might have started the first flicker of life. This was, however, a wasted effort, since a big meteorite several dozen kilometres in diameter hit the Earth. This cataclysm sterilised the Earth by vaporising all its water. Everything had to start again from scratch.

So, to estimate how life appeared on Earth, you need, among other things, to know when this meteoritic shower ended. But the clues are few. Our planet is not at all unchanging. Whether it's volcanism, erosion or continental drift, it's perpetually transforming itself, so that nothing remains of its ancient facial features. Luckily for the experts, the Moon, the Earth's faithful companion, has been dead and frozen for billions of years, and still shows its scars from that tumultuous epoch. Today it's estimated, using the study of lunar craters, that the great sterilising impacts must have stopped between about 4 and 3.8 billion years ago. This fits nicely with the oldest traces of life on Earth which date to about 3.85 billion years ago.

These oldest traces are not fossils, but rather indirect clues left here and there by a form of life of which we know nothing at all. We find these in metamorphic rocks (sedimentary rocks that have been reworked by tectonic movements) found in the southwest of Greenland, which are about 3.85 billion years old and which have a high concentration of carbon-12, the stable isotope of carbon that life prefers to carbon-13, which is as stable but has an additional neutron. The carbon-12 signature in these rocks makes us think that they once contained organic carbon, derived from biological activity. But this is only an hypothesis, and there are nonbiological means that could explain the isotope's presence in the rock.

The first indisputable traces of life on Earth date back 3.5 billion years. On the one hand there are single-celled organisms, filamentary cyanobacteria, whose shape was imprinted in sedimentary rocks (in some cases, there are even filaments of DNA, the long molecule that contains the genetic heritage of a cell). On the other hand, there are stromatolites, structures shaped like big mushrooms, built up layer by layer by colonies of bacteria in shallow water. The oldest stromatolites are found in Swaziland and in Western Australia and are dated at 3.3–3.5 billion years old.

Let's do our sums. The approximate end of the cataclysmic rains of meteorites was about 4 billion years ago. The approximate beginning of life on Earth was about 3.9 billion years ago. Thus difference between these is 100 million years, but to be generous, let's say there was 200 million years in which life was able not only to appear, but also to maintain itself and to prosper. Congratulations to life! Not only is this a ridiculously short time delay compared to geological time scales, but also was carried out excellently. But how did it start? Researchers dream of finding out. It's here that snags turn up and we come across the classical paradox of the chicken and the egg.

First let's give out the starring roles in the film of the primordial cell. Firstly, as the maintenance workers of cellular life, those that guarantee the functioning of an organism, there are the proteins, whose building bricks are the amino acids, twenty of which are essential to life. Among the proteins, there are the enzymes, whose task is to set off the biochemical reactions that are crucial for life. Secondly, there are the nucleic acids, like the indispensable DNA (DeoxyriboNucleic Acid) and the just as essential RNA (RiboNucleic Acid). It's on these long filaments, the characteristic double helix structure of which was discovered by Francis Crick and Jim Watson in 1953, that all the data regarding our genetic identities are stored. DNA's role is clearly crucial since it's this that serves as a genetic messenger from one generation to another. As everything always starts with the splitting of a cell, its double helix structure perfectly fits the role. It's able to separate into two like a zipper. The two separate strands each carry half the information. The mother cell keeps one strand, from which it will construct, using the proteins, a new double helix, and gives the other to its daughter, which also reconstitutes the missing half.

Simplifying a bit, we can present DNA as a long chain composed of four different elements, called bases: adenine (A), thymine (T), cytosine (C) and guanine (G) (in RNA, uracil replaces thymine). Each rung of the double helix molecular ladder is composed of a junction of two bases. If, on one strand, you find a guanine, you can be sure that it's linked by an adenine to the other strand. Just as exclusive matches occur between cytosine and thymine. As a result, when the double helix is divided, the corresponding strand can be reconstructed simply by respecting these molecular matching rules.

All life on Earth, from the smallest cell to the largest animal, is based on this intimate association between nucleic acids and proteins.

There's nothing alive that doesn't have its DNA filaments, nothing for which the choice of bases is different. From this we can conclude that all of today's life forms are derived from a primordial organism that the experts nickname LUCA, an acronym for 'Last Universal Common Ancestor'. We don't know how its different constituents were formed, neither the amino acids that are essential for making proteins, nor the bases that form nucleic acids when attached to a sugar and a phosphate group.

It's common to talk of a 'primordial soup' when referring to the prebiotic period. In a stagnant pool, organic molecules would have reacted with one another to eventually construct the building blocks of life and give birth to the first cell. This scenario is not far from that imagined by the Russian biochemist Alexander Ivanovitch Oparin in 1924. He described a primitive atmosphere different to that of today. Necessarily, he reminded us, in the absence of life, photosynthesis will not yet have had the time to produce much of the atmospheric oxygen. Instead, there would have been, in particular, methane, ammonia, water vapour and a little hydrogen: this would have been a typical reducing atmosphere.[1]

In 1929, the British chemist John Haldane perfected the theory. He imagined different complex molecules made on the surface which fall into the oceans in order to make a sort of hot soup, a thick prebiotic stew where molecules of greater or lesser complexity assemble and finally give birth to life. But for this assembly to take place, there has to be a source of energy, a force that stirs the soup and allows its components to mix. Haldane and Oparin thought this energy came from the Sun's ultraviolet rays or from lightning.

At the beginning of the 1950s, two men decided to submit this scenario to experiment: Harold Urey, a professor at the University of Chicago, and Stanley Miller, a student. They sealed a chemical mix resembling Oparin's primitive atmosphere in an apparatus made of tubes

---

[1] A reducing atmosphere is typically an atmosphere with a lot of hydrogen and hydrogen compounds such as ammonia, which give electrons (negatively charged particles) to other atoms in chemical reactions, and hence 'reduce' the electrical charges of other atoms (make them more negative).

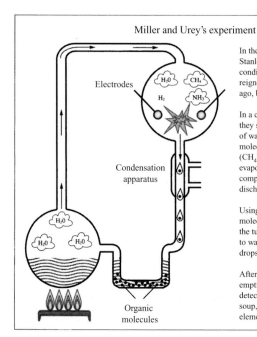

Miller and Urey's experiment

Electrodes

Condensation apparatus

Organic molecules

In their experiment, Harold Urey and Stanley Miller tried to reproduce the conditions that, according to them, reigned on Earth about 4 billion years ago, before life appeared.

In a circuit made of tubes and flasks, they sealed an atmosphere composed of water (H_2O), ammonia (NH_3), molecular hydrogen (H_2) and methane (CH_4). When heated, the water evaporates and joins the other components in a bulb where electric discharges are induced.

Using this energy source, complex molecules form and descend along the tube to a recovery vessel, thanks to water vapour transformed into fine drops by the condensation apparatus.

After several days, Miller and Urey emptied the recovery vessel and detected the presence, in this brownish soup, of several important prebiotic elements such as the amino acids.

and glass bulbs. In one of the flasks, water was brought to the boil. The steam mixed with the other elements and transported them to another flask where electric discharges simulated primordial lightning. The two researchers let their system run in a loop for a week before opening it and analysing the contents. From the viscous, brownish soup they extracted several amino acids, some of which are involved in the construction of proteins. The experiment was reproduced, improved, run in various ways. As long as the atmosphere was a reducing one, as described by Oparin, biologically interesting molecules were produced in large numbers. These molecules even included the components of DNA with the exception of thymine.

Yet, the experts were less and less convinced that the atmosphere had been as reducing as that imagined by Oparin. Several clues pleaded in favour of air saturated with carbon dioxide like on Mars and Venus. This detail considerably changed the game plan. In such an environment, Urey and Miller's experiment produced nearly none of the complex molecules that life finds tasty. So new theories emerged that appeal either to the sky, or to the oceans.

ARRIVALS FROM SPACE?

Complex molecules are not uniquely terrestrial. We now know that they populate the Cosmos. Radio astronomers who have delved into interstellar clouds have discovered more than eighty types of organic molecules in these great clouds. Chemistry is universal. The carbon type meteorites found on Earth abound with vital organic compounds: alcohols, urea, amines, amino acids, bases. The meteorite that fell in the Murchison region (Australia) in 1969 showed the presence of seventy amino acids of which eight are used in the construction of proteins, as well as three bases, adenine, guanine and uracil. In 1986, the European probe Giotto flirted with Halley's Comet, which enabled it to show that the comet is composed of about one quarter organic material. Another European mission, named Rosetta, will study the Wirtanen Comet in 2011. Thanks to their icy nuclei, comets seem not only to have provided the Earth with the water necessary for the development of life, but they could have also enriched this water with complex molecules.

Today, the Earth still daily receives several dozen tonnes of extraterrestrial matter from space in the form of tiny grains that are no larger than a half millimetre. This dust can be found in the polar ices – one of the great experts in the field is the Frenchman Michel Maurette – and it mostly turns out to be composed of particles of carbonaceous chondrites. Was the Earth fertilised from space? More and more scientists are tempted by the idea. Some of them even go further and estimate that not only could the building blocks of life have come from space, but life itself could have travelled here. This is the panspermia theory, which claims that entire organisms were born on other planets before being thrown out into space following some cataclysm, drifting around and landing on a bubbling Earth. This school of thought has existed since the middle of the nineteenth century: a German by the name of Richter mentioned the possibility that the seeds of extraterrestrial life had fertilised the Earth by riding meteorites.

In the early twentieth century, the Swede Svante Arrhenius (1859–1927), a Nobel Prize laureate in chemistry, filed away his German colleague's meteorites in a drawer and proposed a scenario in which stellar photons were able to push microorganisms, spores, for example, across space, since spores are able to hibernate for hundreds of years and resist all sorts of destructive agents. Arrhenius even calculated the time that such organisms would take to travel from the nearest star, Alpha Centaurus, to the Earth: 9000 years. The Swede's theory got stuck at a major objection: while spores might be robust, these microorganisms could not, without protection, escape the aggressive destruction of ultraviolet rays.

Today, this school persists. Its disciples generally estimate that the time available for life to appear on Earth, the 100–200 million years that separate the end of the destructive meteorite downpour from the appearance of the first signs of biological activity, is simply too short. On Mars, however, because the red planet is less massive and since its gravitational attraction is weaker than that of the Earth, the cataclysmic phase would have stopped earlier. At the time, it is supposed, Mars had a denser atmosphere, a higher temperature and liquid water. In short, it would have united the necessary conditions for the creation of life earlier than its cousin, the Earth. Single-celled Martians could have developed on its surface before being thrown out into space to come and fertilise the blue planet.

This vision endows its travelling organisms with an extreme durability. They not only have to survive the impact that threw them off Mars and the destructive rays that traverse space, but also the entry into the terrestrial atmosphere and landing. Isn't this all a bit much, even for resistant single-celled creatures? There are some indications that the feat is not impossible. The Stone experiment, jointly run by the Centre National d'Études Spatiales and the European Space Agency, is, for example, trying to verify the resistance of different rocks, sedimentary and volcanic, to entry into the terrestrial atmosphere. To do this, they placed mineral samples on the shield of the automatic Russian satellite Foton that was destined to come back to Earth. The good news was that a sedimentary rock, of the sort that could hold Martian organisms, survived the return journey. As a second step, an experiment should be staged to try to repeat this with other samples into which microorganisms have been injected.

A MARINE EVENT

Instead of appealing to the sky as the source, another theory prefers to bet on the oceans. This scenario has two major advantages. Buried deep beneath the seas, essential components of life could have been spared from the meteoritic downpour that rained on the surface and could have thus had a much longer period to find the formula for life. Secondly, by evolving in their oceanic hiding places, the same components would have escaped the aggression of the more energetic solar rays like the ultraviolet ones, which we know are dangerous for nucleic acids like DNA.

The energy question remained. Without energy, there's no way to convince the bricks of life to get together. This is no longer a problem, or nearly so, since the discovery of hydrothermal sources in the deepest oceans. These deep-sea events spit out water at temperatures that can reach 450 °C. Near them an incredibly diverse fauna of fish, crustaceans and molluscs has developed. This is an incredible food chain of which the first link is constituted of billions of bacteria that get their energy from the sulphuric compounds ejected by the bubbling water, thus accomplishing a sort of chemosynthesis.

Ocean beds have often been the site of significant magmatic activity. Rocks, criss-crossed by numerous cracks, allow the infiltration of water right up to the edges of the lava. On contact, the liquid is heated, sometimes up to 1000 °C, and then ejected by these deep-sea chimneys. The ejected water is enriched with different minerals; once mixed with the sea water, these minerals participate in the formation of organic molecules which can then combine at leisure to make life.

These three models of the origin of life on Earth are not mutually exclusive. You can imagine an Earth fertilised by dust that comes from space and which then falls to the depths of the oceans, close to the hydrothermal sources. But whatever option we choose, we still have to face a major question, that which punctuates the question of the properties of the first living organism.

THE CHICKEN OR THE EGG?

So we're back again to our good old problem of the chicken and the egg, which we now have to state clearly: which came first, the proteins or the nucleic acids? Suppose that the proteins came first. They have all the required properties except that of autoreplication. Everything they do is done according to the great DNA book, the guardian of the genetic code. So proteins are out of the race, leaving the hypothesis of a primordial nucleic acid. But here again, there's a major objection. DNA may well contain all the necessary information for its repro- duction, but it can't do anything without an intermediary to catalyse the process, a role which is normally carried out by certain proteins. In short, the original organism had to assemble a protein capable of reproducing itself or a nucleic acid with the talents of a protein.

Many experts favour the latter hypothesis, i.e. a nucleic acid with catalysing skills. It would not be a DNA, but an RNA, which, in today's cells, carries out certain specialised tasks. It was in the 1960s that some researchers – Carl Woese, of the University of Illinois, Francis Crick in Great Britain and Leslie Orgel at San Diego – proposed this solution after having noticed that the components of RNA are easier to make than those of DNA. Twenty years later, in 1983, Thomas Cech of the University of Colorado and Sydney Altman dis- covered the first ribozymes, i.e. enzymes made of RNA capable of catalysing chemical reactions. This supported the primordial RNA hypothesis, even if several points remained unresolved, like that of the assembly of this first nucleic acid, its sugars, its phosphates and its bases.

Are we the children of primordial RNA? Maybe. To say more, we would have to find a form of primitive life elsewhere than on Earth.

It would then be like a window on our beginnings. We'll have to be patient. Even if the closest star, Alpha Centaurus, hosted a habitable and inhabited planet, it's 4.3 light-years from the Sun. And at present our technology is still far from allowing us to undertake such a long journey. For the moment antimatter motors only exist in science fiction.

LIFE IN THE SOLAR SYSTEM

With no access to another planetary system, we can profit from examining on our own, since it's not impossible that life could have existed or still exists on one of the Earth's neighbouring planets.

Mars, our red cousin, has long been the planet most suspected of harbouring life. We were reminded in Chapter 2 how the amateur astronomer Percival Lowell, at the end of the nineteenth century, was sure that Mars was inhabited by a decaying civilisation. It took until the 1960s and 1970s to take a closer look. The various American Mariner probes provided good first glimpses of the terrain and ended any final hopes of finding an oasis of life on Mars. In the south, the surface looks uncannily like the Moon, dotted with impact craters. In the north, it hosts huge extinct volcanoes, like Mount Olympus, which peaks at a height of 25 000 metres. But there is no sign of civilisation: no highways, no cities, no football fields and a distressing lack of liquid water. An atmospheric pressure that is a hundred times lower than that of the Earth, means water can only exist in the form of cold steam or ice. It's difficult to imagine how life could be satisfied.

However, liquid water may have existed on Mars until about 500 million years ago. Mariner-9 showed the existence of long structures resembling dried up river beds. It's even possible there was an epoch when a huge ocean covered a large part of the northern hemisphere. The red planet must have once had higher temperatures and pressures than it does today, as well as a larger atmosphere due to the active volcanism. The atmosphere has to have been mainly composed of carbon dioxide, a very efficient greenhouse gas, capable of warming the planet's surface. In short, there is nothing to prevent us believing that in the distant past, the conditions needed for the appearance of a simple and single-celled form of life could have existed on Mars.

In 1976, the *Viking* missions hoped to find some traces of life. If there's still any life on Mars, it must have buried itself in the ground in order to escape the destructive solar rays and colonised the water crystals that may still be there. After digging up samples from the Martian soil, the probes carried out three tests. The surprise was that they were all positive. In each case, the instruments measured effects compatible with a biological origin. Despite this, NASA experts recommended great caution, for two reasons. Firstly, all the phenomena observed during the experiments could also be the result of purely chemical activity. Secondly, the mass spectrometer on the probe which was capable of identifying organic components didn't detect any trace of carbon-based molecules.

For twenty years, the enthusiasm for life on Mars withered in the scientific community. It was left to the authors of science fiction and to flying saucer fans. But a new shiver ran through the community in August 1996, when a team of NASA researchers led by David McKay announced at a press conference that they had discovered possible microscopic fossils in a Martian meteorite, named ALH 84001. Discovered in 1984 buried in the Antarctic ice, where it had rested for about 10 000 years, this rock had to wait a decade for its origin to be formally established by the American David Mittlefehldt. Compared with the other fifteen or so known Martian meteorites, ALH 84001 is unusual because of both its age, at least 4.5 billion years old, and its unusually high concentration of carbonates. On the Earth, carbonates are often linked to biological activity. Also, these carbonates contained special organic molecules, PAHs (Polycyclic Aromatic Hydrocarbons), which can be produced by the decomposition of biological material. Another seductive clue is the presence of tiny magnetite crystals, the very pure form of which is reminiscent of those produced by certain terrestrial microorganisms that orient themselves in line with the Earth's magnetic field.

As thought-provoking as they are, these clues were not conclusive. They could all have been the results of purely chemical reactions. One of the important points concerns the temperature at which the carbonates formed. David McKay and his team suggest temperatures of about 100 °C, which is compatible with the maintenance of life. However, other geochemists favour higher, sterilising temperatures, of the order of 250–500 °C. Meanwhile, critics suggested terrestrial contamination as the source of the PAHs: it was suggested that, on melting, some of the ice around the meteorite could have infiltrated right into the core, depositing the molecules in high concentrations.

Just as contested are the photographs taken by the American team using a powerful electron microscope. You can see tiny structures in the form of worms. It is possible these are the remains of fossilised bacteria no bigger than 0.1–0.2 micrometres long and 0.03 micrometres wide. However, for the sceptics, these tiny rectangular forms are probably an effect of mineral disaggregation. Others doubt that life is possible in such small envelopes.

In short, the ball is still up in the air. A good way to decide the issue would be to find one of the Martian bacteria fixed midway through cellular division. It might also be possible to take one of the structures and cut it into two, looking for subtleties that would show the boundary between a membrane and a cellular interior, but this would be a difficult challenge for even the most powerful microscopes.

Whatever the truth about ALH 84001 is, scientists continue to believe in the possibility of ancient life on the red planet. In 2003, two probes will touch down on Martian soil. The first, Mars Express, consists of an orbital probe funded by the European Space Agency and a landing module called Beagle 2 – the *Beagle* was the ship which took Charles Darwin to the Antipodes. Beagle 2 is a mainly British project run by Colin Pillinger from the Open University. This rigid, light probe (it won't be above 30 kilograms) will carry on board a very powerful mass spectrometer that should be able to test rocks, soil and the atmosphere, looking for clues to past and present life on Mars.

Mars Rover 2003, an American project from NASA and the Jet Propulsion Laboratory, is heavier and at least ten times more expensive. It includes two mobile robots, inspired from little Rocky of the 1997 Pathfinder mission, which will weigh (on Earth) about

150 kilograms each. On their six wheels, they'll be able move up to 100 metres per day. Each robot will be released in a special region where it will primarily search for geological and other clues to Mars' past climatic history in order to learn more about the past presence of liquid water, since without water, there's no life.

But no robot can replace a human and the vast array of scientific instruments that he has available on Earth. A clue to life could be missed by the probes. There's less risk that it would be missed by laboratory analysis. This is why the experts are planning further missions to Mars in order to bring back Martian rock and soil samples. These expeditions should take place between 2007 and 2010. And then? Then, humans will probably walk on Mars, possibly between 2020 and 2030, a project so difficult that it cannot happen unless it's an international effort.

Mars is not the exobiologists' only target in the Solar System, and it seems more and more probable that life could have tried its luck elsewhere. This astonishing perspective owes a lot to the discovery on the Earth of organisms living in particularly difficult ecological niches. A number of species of bacteria, called extremophiles due to their ability to survive in extreme environments, have made a strong impression on scientists. We have already seen the hyperthermophilic organisms that live close to boiling hot deep-sea hydrothermal sources. Other microbes, the psychrophiles, prefer the cold and willingly colonise polar waters and permafrost. In August 2000, the coldest and hottest records for extreme temperatures for life were respectively $-17\,°C$ and $+114\,°C$.

We should also mention the acidophilic bacteria that live in very acidic media (pH < 3) and the alkalinophiles who feel comfortable in very alkaline environments (pH > 10). Both have the peculiarity of maintaining a neutral pH inside the cell using enzymes placed near the membrane, extremozymes, which act as a barrier to these very aggressive media, which are especially toxic for structures as fragile as nucleic acids. In 1997 Tullis Onstott's team from the University of Princeton also discovered bacteria living at depths of around

3500 metres under the ground. These strange creatures populate the cracks between mineral grains and while one species feeds on inorganic chemical compounds from its environment, another lives off the organic materials provided by the first.

These extremophiles offer new perspectives to exobiologists. If life can cope with such extreme conditions, maybe it exists elsewhere in the Solar System: on Venus, for example, the Earth's twin sister by size. But all the same, its unlikely. Its closeness to the Sun and its greenhouse effect guarantee an average temperature of 450 °C, while biologists think that the limit for an extremophilic organism must be somewhere around 150 °C. In contrast, it's possible that in the long distant past, Venus tried to play with the gift of life when its atmosphere contained more water than the tiny amounts detected today and when the Sun, younger but already in its stable phase, was about 30% less luminous than today.

If Venus isn't an interesting target for exobiology, where should we turn? Surely not towards Jupiter or Saturn? It's difficult to imagine that life could take root in these giant gas worlds, but we can consider their satellites. Around Jupiter, astronomers are especially optimistic about the satellite Europa, a small ball entirely covered with ice. If scientists are ready to believe that life exists on this frozen world, it's because they also have their eyes on another jovian moon, Io. Slightly larger than Europa, it's also closer to Jupiter. Io is a world of sulphur and fire, of volcanoes and fury which spit their anger up to 30 kilometres high. The origin of this volcanism is to be found in the huge tidal effects induced by the mother planet, which distort the small satellite. This constant distortion forced Io to become very actively volcanic. Even though little Europa – which is smaller than our Moon – is further from Jupiter than Io, it too must suffer from the tidal effects of its mother planet.

How does it cope? That's the key question. You can imagine that under the ice, in its depths, Europa too is volcanically active. You can even imagine that this contributes to the existence of a large layer of liquid water. We would then have, like at the bottom of our oceans, hydrothermal sources that could provide energy and the essential building blocks for the appearance and the maintenance of some form of life. There are indications that make us think that under Europa's ice, liquid water really does exist. Many photographs (of which the latest were taken in 1998 by the Galileo probe) show an ice surface covered by numerous cracks filled by what seems to be newly formed water ice. But the question of the depth at which water can be found remains open. It could be 200 metres, or a few dozen kilometres. The experts would prefer the former. These are the experts who some day, when technology will allow it, hope to send a mission to Europa to dig into its ice cover and discover what's hiding there.

It was once thought that Titan, Saturn's biggest moon, might be able to host life. The absence of water on its surface is fatal for this hypothesis. There are lakes, but these are lakes of ethane and methane. However, this object fascinates exobiologists because in its atmosphere, which is as dense as ours, they see a reducing atmosphere similar to that tested by Miller and Urey in 1953. So it would be interesting to see if complex, prebiotic molecules form there. In any case, this is what the European probe Huygens, coordinated by the French exobiologist François Raulin, will try to detect. Carried on the American vessel Cassini, the Huygens probe should reach its objective in 2004.

HABITABLE PLANETS

Thanks to the scientific developments of the last forty years, researchers have started to gain some idea of what conditions seem to be necessary for life to start. There's hardly any question that liquid water is crucial. It's a premium quality solvent, able to dissolve most biochemical molecules and to accelerate reactions between various components. And to have liquid water, you need to have the right temperature. This can be provided by underwater volcanism, but it's likely that in that case the biological niche would be relatively constrained. In contrast, the niche is much bigger if the ground temperature is favourable.

The existence of liquid water on the surface of a planet depends on the distance between the planet and its star. It can't be either too far or too close. The Earth lies at 150 million kilometres from the Sun, i.e. an astronomical unit. According to the experts, the habitable zone around our star extends from 0.95 astronomical units to 1.5 astronomical units, a range that varies as a function of the Sun's brightness, which, of course, changes throughout its life.

Around a star twenty-five times brighter than our own, a twin planet of the Earth would have to be located five times further out than the Earth. Around a star ten times fainter than the Sun, the twin would have to be in an orbit corresponding to that of Mercury. It's what the experts call the golden orbit. Clearly there are limiting cases. Around a very faint star, for example, the golden orbit would be located inside the tidal zone, forcing the planet to always show the same face towards its star. With one face constantly lit and the other shrouded in unending night, life might not be able to cope.

Distance is not the only consideration. The planet also has to have enough mass to retain sufficient atmosphere . Our Moon cannot. All the same, the same Moon might be an essential element in the appearance of life on Earth, if we give credence to the ideas of Jacques Laskar and Philippe Robutel, of the Bureau des Longitudes (Longitude Institute) in Paris. These two physicists calculated that without the stabilising effect of our satellite via the tidal effect, the Earth's axis of rotation would be tilted every now and then, up to an angle of about sixty degrees, which would make equatorial regions slide to the poles and vice versa. Such drastic changes would have radical climatic consequences to which it's not sure that life could have adapted.

As well as the Moon, Jupiter would also seem to play its part in the development of life on Earth. It was the American theorist George Wetherill who first proposed the theory according to which our largest great gas giant is a sort of shield for the Solar System. Its significant gravitational attraction – remember that its mass is about 318 Earths – would have caught most of the comets that came close to it. Without it, the Earth would have experienced many more cometary impacts throughout its long history: there would have been a great cataclysm every hundred thousand years or so. Life on Earth probably wouldn't have had the elbow room needed to go beyond the single-celled stage.

So terrestrial life, with its extraordinary complexity and variety, has been able to count on a fantastic ally: stability. A moon to keep the planet in place, a Jupiter to get rid of cometary intruders, and also a circular orbit that lets it always stay at the same distance from the Sun. If our planet had followed an even slightly elliptical path, the continual changes in distance from the Sun would have induced very strong and possibly even uninhabitable seasonal extremes. Sure, its atmosphere, thanks to air currents, would have been able to reduce the differences, but nothing says that this would have been enough to make the place habitable.

Since we're talking about the atmosphere, let's examine this further. Today, the Earth's atmosphere contains about 21% oxygen. The vast majority of this (at least 99%) was produced by photosynthesis. But the microorganisms and the plants which use this breathing method have a great need for carbon dioxide in order to survive. Even if the Earth had large quantities of carbon dioxide available during its youth, it had nevertheless to find a way to continually produce it, so that it didn't become dangerously scarce. Its manufacturing secret is active volcanism, the result of plate tectonics, which guarantees a continuous production of carbon dioxide from its gas emissions.

If we limit ourselves to terrestrial orthodoxy and to the reasons that make our planet so fertile for life, we can assert that none of the exoplanets discovered so far satisfies these conditions. In contrast, it's possible that one of the gas giants, which is neither too close nor too far from its star, and whose orbit is not too eccentric, could host a satellite whose surface is bathed in liquid water. Maybe an attempt at life occurred there. And as for the question of highly evolved life living there...

ON THE TRAIL OF INTELLIGENCE
It's more than forty years since people have tried to detect extraterrestrial signals. Is this an outlandish hope of romantic dreamers? Maybe

a bit, but it's also the hope of rigorous scientists. It all started at the end of the 1950s with Frank Drake, a brilliant American student at Harvard. At the time, he was participating in the early days of radio astronomy, the heir of radar developed in the Second World War, and soon found himself at the Green Bank Observatory, in Virginia. His career path was classical, his thought less so. He was convinced that radio telescopes could be used to detect radio signals sent by extraterrestrial civilisations, but fearing his colleagues' mockery, he preferred to keep quiet. Then in 1959, the review *Nature* published an article by two eminent scientists, Philip Morrison and Giuseppe Cocconi, entitled 'In search of interstellar communication', which urged the scientific community to commit itself to the search. Relieved to discover that he was not alone in thinking about little green men, the young student decided to talk about his passion to his director, Otto Struve. To the great surprise of the young physicist, the latter was enthusiastic.

On 8 April 1960, Frank Drake aimed the 85-foot Green Bank antenna at its first target, the star Tau Ceti. Hours passed without any peculiar signal troubling the calm of the place. Drake knew that his quest was to be long and difficult, but he only had a limited amount of observing time. His modest project, named Ozma in allusion to the story of the wizard of Oz, was intended to last only a year. The young American then turned his telescope towards its second target, the star Epsilon Eridani (around which we've now discovered a planet). The antenna had barely stopped when a heavy and regular signal was heard.

The young radio astronomer was thunderstruck. Could it be that easy? Could chance have smiled on him to such an extent? He had to calm down and apply the procedure planned for such an event, which is to shift the telescope and then come back to the target. If the signal is still there, then it's real. However once it got back to the right position, the antenna no longer picked up any signal. Ten days later, the signal showed up again; however, this time, Frank Drake had prepared some equipment for showing the difference between a genuinely stellar signal and a terrestrial signal. This apparatus showed that the signal was a terrestrial parasite, or more precisely, it was the passage of high altitude American spy planes, the famous U2s.

A year later, project Ozma was over, without success, but the hunt for extraterrestrial signals didn't stop there. The first conference on the subject was held, with several prestigious participants, and participants destined to become prestigious, like Carl Sagan. It was at that meeting that Frank Drake presented the famous equation that now carries his name:

$$N = R f_p n_e f_l f_i f_c L.$$

This sequence of letters has become a great classic in the search for extraterrestrial signals. It makes it possible to estimate the likely number of intelligent civilisations in space, depending on the following parameters:

N: the number of civilisations that exist in our Galaxy
R: the star formation rate in our Galaxy
f_p: the fraction of these stars that have **planets**
n_e: the number of habitable (**earthlike**) planets per solar system
f_l: the fraction of these planets that have developed **life**
f_i: the fraction of these that have developed **intelligence**
f_c: the fraction of civilisations that **communicate**
L: the lifetime of a civilisation

The result is radically different depending on whether you're an optimist or a pessimist. So, some claim that thousands of advanced and communicating civilisations populate our Galaxy. Others, claim that we would have to go through several millions of galaxies in order to have a tiny chance of hearing from advanced extraterrestrials with as much curiosity as us.

In this equation, some of the factors are more or less known: R, for example, the star formation rate in the galaxy, has an average value of 1, i.e., one birth per year; f_p can also be approximated, using the results provided by radial velocity detection. We know that about 5% of solar type stars have planets. Nevertheless, this value is biased since the radial velocity method can only detect massive companions close to their star. It says nothing about lower mass planets, which could well be much more numerous. Supporting this hypothesis is the fact that all young stars show accretion discs, which are destined to change into protoplanetary discs. Physics is the same everywhere. There's no reason for the planet frequency not to go above 5%. We could reasonably raise this figure to 50%.

The other values are mere speculation. Are habitable planets frequent? Is life an exceptional event or is it a sort of virtually automatic consequence of prebiotic chemistry in a favourable environment? Does the process of natural selection forcibly tend towards the direction of intelligent life forms? And if supposing we say 'Yes' to this, would these sentient beings necessarily share with humanity this profound curiosity for the mysteries of Nature?

All in all, Frank Drake's equation – and its author is the first to admit this – doesn't really have any scientific value. It's not much more than a way of getting started on discussions of extraterrestrial life. If an answer is found, it'll come from experiment.

Those who've been searching for extraterrestrial signals for more than forty years are indisputably patient people of the optimist camp. They still haven't found anything and yet they untiringly continue their painstaking task. Not only are there billions and billions of stars, but also, for each star, there are billions and billions of possible transmission channels. While Frank Drake's pioneering efforts were particularly fastidious, technical advances make it possible today to simultaneously search over several million channels. An apparatus cuts the spectrum into narrow bands, of about 1 hertz, because the narrower a band is, the higher the chance is that a signal in it is artificial. Typically, the organisers of this quest favour a frequency domain located between 0.5 and 50 gigahertz, because that's where cosmic background noise interferes the least, a detail that would undoubtedly have been noticed by Galactic civilisations experienced in interstellar communication.

To detect such a signal would be a key moment in the history of humanity. Hoping that it would be deliberate and peaceful, a response would be in order. But, unless you can think of some exotic physics, an intergalactic conversation would be boring to say the least. This is because, as we've known since Einstein, electromagnetic waves can't travel faster than the speed of light, i.e. 300 000 kilometres per second. On our scale, that's extremely fast, but when you want to transmit to a planet 100 or 200 light-years away, there's the problem of slowness. Even just exchanging greetings would take more than a thousand years.

Rather than exchanging a deliberate message, it's possible that we could pick up more complex radio emissions from a close exoplanet. On the Earth, it's about seventy years since we started emitting radio waves, of which some have spread into space starting journeys that continue to this day. We are surrounded by a sort of communication bubble with a radius of about 70 light-years. Maybe a society which has been communicating much longer than us has already included us in its communication bubble. But it would not be easy to detect. If they're not deliberate attempts at communication, these signals would probably be amplified very little at the source, so they would be very faint.

Whatever the chances of success may be, the SETI people are continuing their project. In the USA, where they are becoming more and more numerous, for many years they were able to rely on public funding, possibly because people in the Soviet Union were also very active in this domain. In 1993 the American Congress decided to cut the supply line to the SETI programmes. Since then, they survive thanks to the generosity of private donors.

This funding enables them to follow several strategies. The most focussed of these is called project Phoenix. Managed by the SETI Institute in Mountain View, this experiment has used, since 1998, the biggest non-interferometric telescope in the world, that of Arecibo on the island of Puerto Rico. The organisers of project Phoenix take advantage of this 305-metre diameter stationary antenna supported in a dip in the ground – it's the secondary antenna placed 130 metres higher that is moved – to examine 1000 solar type stars, that are at least 3 billion years old and are located within 200 light-years of the Sun.

The Serendip project has a completely different philosophy. While it too uses the Arecibo telescope, it does it virtually without other users noticing. It's like a sprig of mistletoe feeding off the highest branch of a tree. In fact, the Serendip project receiver is placed on the secondary antenna of the telescope. The advantage of this is that it doesn't interfere with normal observations. It's happy to be carried around according to the wishes of those who've been awarded observing time for other research projects. It listens, scrutinises, observes wherever it's led and is happy. As it nearly always has its nose pointed towards the sky, this receiver records a really astronomical amount of data. This is its strength, but also its weakness. Because if it registers an interesting signal, it's not allowed to immediately verify its source, nor even on the following day. And that's not all. The mountains of data recorded need to be treated and analysed by computers. The computing power required for such a task is mind-blowing. Now that the Internet is more widely available, SETI@home gives cybernauts the opportunity to download a small programme that makes it possible to use their computers whenever they're not being used otherwise. The home computer analyses the data from Serendip that the user has obtained via the Internet. Once the decoding work is over, the home computer sends the reduced data back, and picks up another packet of raw data. Launched in May 1999, this project has, according to its organisers, met with resounding success. After more than four years of existence, SETI@home's team said it relied on the computing support of more than four million home computers around the entire world. So if we detected an extraterrestrial signal thanks to SETI@home, it would be a world discovery in more ways than one, and it might be time to learn a new language, Extraterrestrian.

Appendix. Properties of the exoplanets

Discovered as of 7 October 2002

Planet	Mass (in Jupiter mass)	Period (in Earth days)	Eccentricity
HD 49674 b	0.12	4.95	0
HD 76700	0.197	3.971	0
55 Cnc c (HD 75732 c)	0.21	44.28	0.34
HD 16141 b	0.22	75.8	0.28
HD 168746 b	0.23	6.403	0.08
HD 46375 b	0.25	3.024	0.02
HD 83443 b	0.34	2.985	0.08
HD 108147 b	0.4	10.901	0.498
HD 75289 b	0.42	3.5097	0
51 Peg b (HD 217014 b)	0.47	4.229	0
BD-10 3166 b	0.48	3.487	0.05
HD 6434 b	0.48	22.09	0.3
HD 187123 b	0.54	3.097	0.01
GJ 876 c	0.56	30.12	0.27
Ups And b (HD 9826)	0.68	4.617	0.02
HD 209458 b	0.685	3.524 33	0
47 UMa c (HD 95128 c)	0.76	2594	0.1
HD 38529 b	0.77	14.31	0.27
HD 4208 b	0.81	828.95	0.04
HD 82943 c	0.88	221.6	0.54

(*cont.*)

Planet	Mass (in Jupiter mass)	Period (in Earth days)	Eccentricity
Eps Eri b			
(HD 22049 b)	0.88	2518	0.6
HD 121504 b	0.89	64.62	0.13
HD 114729 b	0.9	1136	0.33
HD 179949 b	0.93	3.092	0
55 Cnc b			
(HD 75732 b)	0.93	14.66	0.03
Rho CrB b			
(HD 143761 b)	0.99	39.81	0.07
HD 114783 b	0.99	500.73	0.1
HD 114386 b	0.99	872.3	0.28
HD 142 b	1	337.11	0.38
HD 150706 b	1	264.9	0.38
HD 147513 b	1.01	540.4	0.52
HD 37124 c	1.01	1942	0.4
HD 130322 b	1.02	10.72	0.044
HD 52265 b	1.05	119.6	0.35
HD 20367 b	1.12	500	0.23
HD 37124 b	1.13	154.8	0.31
Gl 777 Ab			
(HD 190360Ab)	1.15	2614	0
HD 216435 b	1.23	1326	0.14
HD 177830 b	1.24	391	0.4
HD 210277 b	1.24	435.6	0.45
HD 217107 b	1.282	7.1262	0.134
HD 27442 b	1.42	426	0.02
HD 74156 b	1.55	51.61	0.649
HD 12661 c	1.56	1444.5	0.2
HD 134987 b	1.58	260	0.24

(*cont.*)

Planet	Mass (in Jupiter mass)	Period (in Earth days)	Eccentricity
HD 82943 b	1.63	444.6	0.41
HD 4203 b	1.64	406.0	0.53
HD 108874 b	1.65	401.1	0.36
16 Cyg Bb (HD186427 b)	1.68	796.7	0.68
gamma Cep b (HD 222404 b)	1.76	903	0.2
HD 19994 b	1.83	454	0.2
GJ 876 b	1.89	61.02	0.1
HD 68988 b	1.9	6.276	0.15
HD 160691 b	1.99	743	0.62
Ups And c (HD 9826 c)	2.05	241.3	0.24
HD 216437 b	2.05	1119	0.17
HD 8574 b	2.08	228.52	0.304
iota Hor b (HD 17051 b)	2.24	311.3	0.22
47 UMa b (HD 95128 b)	2.54	1089	0.06
HD 23079 b	2.54	627.34	0.06
HD 72659 b	2.55	2185??	0.18??
HD 128311 b	2.63	414	0.21
HD 12661 b	2.84	250.5	0.19
HD 169830 b	2.94	229.9	0.35
HD 73526 b	3.03	186.9	0.41
HD 40979 b	3.32	267.2	0.23
HD 196050 b	3.36	1098	0.22
GJ 3021 b (HD 1237 b)	3.37	133.71	0.511
HD 190228 b	3.44	1112	0.52
HD 195019 b	3.55	18.2	0.01

(*cont.*)

Planet	Mass (in Jupiter mass)	Period (in Earth days)	Eccentricity
HD 92788 b	3.81	340	0.36
HD 80606 b	3.9	111.81	0.927
GJ 86 b (HD 13445 b)	4	15.78	0.046
55 Cnc d (HD 75732 d)	4	5360	0.16
HD 2039 b	4.12	1209.9	0.65
Tau Boo b (HD 120136 b)	4.14	3.313	0.02
14 Her b (HD 145675 b)	4.27	1764	0.353
Ups And d (HD 9826 d)	4.29	1308.5	0.31
HD 213240 b	4.5	951	0.45
HD 50554 b	4.8	1237	0.515
HD 222582 b	5.18	576	0.71
HD 28185 b	5.7	383	0.07
HD 10697 b	6.08	1074	0.11
HD 178911 B b	6.292	71.487	0.1243
HD 89744 b	7.17	256	0.7
HD 106252 b	7.39	1582	0.478
70 Vir b (HD 117176 b)	7.42	116.7	0.4
HD 74156 c	7.5	2300	0.395
HD 168443 b	7.7	58.116	0.529
HD 30177 b	7.95	1620	0.22
HD 23596 b	8	1558	0.314
iota Dra b (HD 137759 b)	8.64	550.65	0.71
HD 141937 b	9.7	653.22	0.41

(*cont.*)

Planet	Mass (in Jupiter mass)	Period (in Earth days)	Eccentricity
HD 33636 b	10	3030	0.56
HD 39091 b	10.37	2115.2	0.62
HD 114762 b	10.96	84.03	0.33
HD 38529 c	11.3	2189	0.34
HD 136118 b	11.9	1209	0.37
HD 162020 b	14.4	8.428 198	0.277
HD 168443 c	16.9	1739.5	0.228
HD 202206 b	17.5	256.03	0.429

Glossary

Absolute magnitude: the intrinsic brightness of a celestial body, i.e. of the amount of light that it really sends into space. This term contrasts with the **apparent magnitude**, which is the brightness of a celestial body as perceived from the Earth. The lower the value of the magnitude, the brighter the celestial body is. Bodies that are as faint as (apparent) magnitude 6 can be seen with the naked eye, while objects as faint as magnitude 9 can be seen with binoculars. As seen from the Earth, the Sun is extremely bright, and so has a very negative (apparent) magnitude, −26.

Accretion disc: a cloud of gas and dust that forms around a star when it is born. Most of the matter is fed to the star, while the rest is transformed into a protoplanetary disc, which as its name indicates, may allow planets to form.

Adaptive optics: a technique that makes it possible to correct the distortions in starlight caused by perturbations in the terrestrial atmosphere.

Aphelion: the point in a body's orbit when it is furthest from the Sun. The **apoastron** is the same thing for a body orbiting around a star other than the Sun.

Arc second: a unit of angular measurement equal to one 3600th of a degree, i.e. less than the apparent thickness of a human hair seen from a distance of about 35 metres. In astrometry, perturbations in the paths of stars caused by planets are much less than an arc second. They are expressed in milli-arc seconds (a thousand times smaller) or micro-arc seconds (a million times smaller).

Asteroid: a small rocky body floating in space. The Solar System includes an asteroid belt located between Jupiter and Mars. It contains well-known objects such as Ceres, Pallas and Vesta, which were discovered at the beginning of the nineteenth century.

Astrometry: the branch of astronomy concerned with the measurement of the positions of objects in the sky. Several things are calculated, including:

(1) The **parallax**, which is the apparent movement of an object in the sky due to the rotation of the Earth around the Sun.

(2) The **proper motion**, which is an object's change in position in the sky year after year while it travels through space.

(3) The **planetary perturbation**, which is an object's residual movement after having calculated and removed the parallax and the proper motion. It is due to the gravitational influence of a body orbitting it. For the detection of exoplanets, astrometry is particularly sensitive to bodies that are relatively far from their stars, while the **radial velocity** technique is more sensitive to planets close to their stars.

Astronomical unit: a widely used unit of length measurement, corresponding to the average distance between the Earth and the Sun, i.e. 150 million kilometres. A **light-year** is just over 63 000 astronomical units.

Binary star: a system of two stars that orbit around one another. Multi-stellar systems are very common. Only a third of stars are solitary like the Sun.

Brown dwarf: an object that forms, like a star, by the fragmentation of an interstellar gas cloud, but is not massive enough to enable the hydrogen fusion required for stars to light up. The threshold below which a star does not light up is theoretically fixed at about 0.08 solar masses. The least massive brown dwarfs should have the same mass as the biggest giant planets, i.e. around 0.01 solar masses. A brown dwarf's surface temperature is generally below 2200 °C.

Centre of mass (barycentre): the point around which two gravitationally linked objects rotate. If the two objects are of equal masses, then the centre of mass lies exactly half-way between them. The more unequal the masses, the closer the centre of mass is to the more massive object. The centre of mass between the Sun and Jupiter is located just outside the Sun's surface. This is enough to cause a slight movement of the Sun, which if detected from a nearby star, would reveal Jupiter's existence.

Comet: an object made of ice and dust that may come from the Kuiper Belt (of which Pluto is the most well-known member), or from the Oort Cloud, a reservoir of small objects located between 40 000 and 100 000 astronomical units from the Sun.

Coronagraph: an instrument that makes it possible to mask the light from a star in order to observe objects very close to it.

Degenerate (matter): a state of matter where the only force able to oppose the force of gravity is a pressure whose nature is particular to quantum mechanics. In **white dwarfs** and **brown dwarfs**, it is electrons that create this state, while in **neutron stars**, it is, as the name indicates, neutrons that create the degenerate state.

Density: the amount of matter divided by its volume. For liquids and solids, the density is often expressed in units of the density of water, while for gases, it is often expressed in units of the density of air.

Deuterium: an isotope of hydrogen in which the nucleus is composed of a proton and a neutron (normal hydrogen only has a **proton**). **Brown dwarfs** are often able to burn deuterium during their very early youth, which makes them brighter and more easily detectable.

Doppler–Fizeau effect: the shift in the spectrum of an object that moves with respect to an observer. This is the effect that causes the sound we hear from the siren of a passing ambulance to change in pitch: as the ambulance approaches, the siren is sharper than its rest pitch; as it recedes, the pitch is flatter. This sound effect also occurs for light. A light source which comes towards us is shifted towards the blue side of the spectrum (to shorter wavelengths), while if the source moves away, the spectrum shifts towards the red (longer wavelengths). When a system including a star and a planet is viewed in profile from the Earth, the light that comes from the star alternately shifts to the red and the blue, revealing the to and fro movement caused by its companion. This is the effect on which the **radial velocity** technique is based.

Eccentricity (orbital): this describes the shape of elliptical orbits. The closer this value is to 1, the longer and thinner the ellipse is. A perfectly circular orbit has an eccentricity of 0. The great majority of exoplanets with **orbital periods** greater than 10 days have higher eccentricities than those typical of Solar System planets.

Electromagnetic spectrum: a continuous scale that includes all types of electromagnetic radiation, from the least (radio waves) to the most energetic (gamma rays). The visible wavelength domain occupies a tiny section more or less in the middle of this scale.

Electron: a negatively charged particle that 'orbits' around the nucleus of an atom.

Ephemeris: a sequence of points that predicts the future positions of a celestial object moving across the sky.

Exobiology: the study of the origins, the distribution and the evolution of life in the Universe.

Exoplanet (or **extrasolar planet**): any planet located outside the Solar System.

Galaxy: a great collection of stars, grouped together in space and held together by gravity. Our own galaxy, also called the Milky Way, hosts more than 100 billion stars.

Gas giant: a planet characterised by the presence of a very large atmosphere, composed mostly of hydrogen and helium, for example, Jupiter and Saturn. It is thought that these planets form by the accretion of **planetesimals** made of ice and dust, until they reach a critical mass (equal to about 10 Earths), which allows them to capture nearby hydrogen and helium gas.

Hydrogen: the first element in the periodic table, hydrogen is also the lightest atom that exists since it consists of just one proton and one electron.

Infrared: a type of radiation whose frequency is just below that of visible light. Infrared makes it possible to detect cold objects that radiate very little in visible light. It's in the infrared domain that the luminosity ratio between a planet and a star is smallest.

Interferometry: a technique for combining the observed radiation beams from two (or more) telescopes separated by some distance. The resolution obtained corresponds to that for a telescope as big as the distance between the telescopes, so that this is like creating a much bigger telescope.

Ion: an atom that has lost or gained one or more electrons.

Kelvin (degree): a unit for measuring 'absolute' temperature. To convert a kelvin temperature to degrees Celsius, subtract 273.15 degrees.

Light-year: the distance that light travels, at the speed of 300 000 kilometres per second in a vacuum, during one year. A light-year equals 0.3067 **parsecs**, which is a unit commonly used in astronomy.

Main sequence: the star family that includes all stars that succeed in starting and maintaining hydrogen fusion.

Microlensing (effect): the process by which the light from a very distant star is amplified by a massive object passing between itself and the Earth. The object curves light rays from the star and concentrates them towards the observer. This is one of the methods used to try to indirectly detect **telluric planets**.

Missing (or dark) matter: matter that has been hypothesised to exist by observing the dynamics of certain **galaxies** like the Milky Way.

Neutron: an uncharged particle found in the nucleus of atoms.

Neutron star: an object that remains after the death of a giant massive star (**supernova**). A typical neutron star contains about one and a half times the mass of the Sun in a diameter of only 20 kilometres. Only black holes have a greater density. Some neutron stars rotate very rapidly and give birth to **pulsars**.

Nuclear fusion: the process by which two light atoms fuse together and in this way release energy. To set off fusion, very special, extreme conditions are required. Hydrogen nuclei only fuse together if the ambient temperature is about 10 million degrees Celsius and if the pressure is about a billion atmospheres. This is why scientists have a lot of trouble trying to control the process in laboratories.

Nulling: a process used in interferometry by which a star can be 'extinguished' in order to look for possible planets close to it.

Parsec: a unit of astronomical length equal to 3.26 light-years.

Perihelion: the point in a body's orbit when it's closest to the Sun. The **periastron** is the same thing for a body orbiting a star other than the Sun.

Period (orbital): the time it takes for an object to complete an orbit. This is often expressed in terms of terrestrial days.

Photon: an elementary particle associated with electromagnetic waves and in particular with visible light.

Planet: an object that forms from the debris left after a star has formed. The debris is distributed in a protoplanetary disc, and an object formed in it is a planet if it attains a large enough mass to be forced by gravity into becoming more or less spherical.

Planetesimal: a small object from a few metres to several kilometres in size, formed from dust and ice contained in a protoplanetary disc. Planets form from the accretion of planetesimals.

Prebiotic chemistry: complex chemistry that allowed the creation of life on Earth and maybe elsewhere.

Proton: a positively charged particle found in the nucleus of atoms.

Pulsar: a neutron star that is rotating very rapidly and that emits flashes in the radio domain of the **electromagnetic spectrum**. These emissions are so regular that atomic clocks are generally required to detect possible irregularities, in particular those due to the presence of a planet orbiting the pulsar.

Radial velocities (method of): a method that uses a **spectrograph** for the indirect detection of invisible objects orbiting around their main stars. The perturbations induced in the main star under the influence of a companion show up as speed changes along our (radial) line of sight. When the star approaches us, its light is shifted towards the blue regions of the spectrum according to the principle of the **Doppler–Fizeau effect**. When it moves away from us, the shift is towards the red. Certain properties of the companion that induces the velocity perturbations can be deduced from the amplitude of these shifts.

Red dwarf: a star that lies at the end of the **main sequence**. Red dwarfs are thus the least massive and the least hot of all stars that are fuelled by stable thermonuclear reactions.

Spectrograph: an instrument used, for example, to photograph the luminous signature (also called a **spectrum**) of a star. Such a spectrum is like a sort of rainbow (continuous spectrum) that ranges from the red to the violet and is punctuated by black lines of various thickness. These lines, call absorption lines, are due to the different chemical elements present in the atmosphere of the star which absorbs at certain very specific wavelengths. By comparing the positions of these absorption lines from a star to the positions of a laboratory spectrum, the spectral shifts, which are so important to the **radial velocity** method, can be measured.

Telluric: describes a dense, solid planet. Mercury, Venus, the Earth and Mars are the four telluric planets in the Solar System. They can be contrasted with the **gas giants** like Jupiter, Saturn, Uranus and Neptune, which are more massive, but less dense, than the telluric planets.

Transit: a phenomenon by which the light seen from a star dims slightly due to a planet passing between it and the observer. The first **exoplanet** discovered by the **radial velocity** method and confirmed by the observation of a transit was HD 209458.

White dwarf: a very **dense** star, of the size of the Earth, but containing a mass about equivalent to that of the Sun.

Bibliography

Some history

Koyré A., *Du Monde Clos à l'Univers Infini*, Paris, Gallimard, coll. 'Tel', 1973.

Tombaugh C.W. and Moore P., *Out of the Darkness: the Planet Pluto*, Harrisburg, Stackpole Books, 1980.

Verdet J.-P., *Une Histoire de l'Astronomie*, Paris, Éditions du Seuil, coll. 'Points Science', 1990.

On how stars and planets are made

Acker A. and Lançon A., *Vie et Mort des Étoiles*, Paris, Flammarion, coll. 'Dominos', 1998.

Goldreich P. and Tremaine S., 'Disk–Satellite interactions', *Astrophysical Journal*, **183**, 1051, 1980.

Lagrange-Henri A.M., Vidal-Madjar A. and Ferlet R., 'The β Pictoris Circumstellar Disk. VI. Evidence for material falling on to the star', *Astronomy and Astrophysics*, **31**, 129, 1993.

Proust D. and Breysacher J., *Les Étoiles*, Paris, Éditions du Seuil, coll. 'Points Science', 1996.

Safronov V.S., *Evolution of the Planetary Cloud and Formation of the Earth and the Planets*, Moscow, Nauka, 1969.

Smith B. and Terrile R.J., 'A circumstellar disk around β Pictoris', *Science*, **226**, 1421, 1984.

A few references on pulsars and planets around pulsars

Hewish A., Bell S., Pilkington J.D.H., Scott P.F. and Collins R.A., 'Observation of a rapidly pulsating radio source', *Nature*, **217**, 709, 1968.

Philips J.A. and Thorsett S.E., 'Planets around pulsars: a review', *Astrophysics and Space Science*, **212**, 91, 1994.

Wolszczan A. and Frail D., 'A planetary system around the millisecond pulsar PSR B1257+12', *Nature*, **355**, 145, 1992.

Some references on brown dwarfs

Basri G., 'La découverte des naines brunes', *Pour la Science*, 272, June 2000.

Chabrier G., 'Entre étoiles et planètes : les naines brunes', *La Recherche*, **290**, 51, 1996.

Kumar S., 'The structure of stars of very low mass', *Astrophysical Journal*, **137**, 1121, 1963.

Martin E.L., Rebolo R., Zapatero-Osorio M.R., 'The discovery of brown dwarfs', *American Scientist*, November–December 1997.

Nakajima T., Oppenheimer B.R., Kulkarni S.R., Golimowski D.A., Matthews K. and Durrance S.T., 'Discovery of a cool brown dwarf', *Nature*, **378**, 463, 1995.

The first planets and planetary systems

Black D.C., 'Completing the copernican revolution: the search for other planetary systems', *Annual Review of Astronomy and Astrophysics*, **33**, 359, 1995.

Butler R.P. and Marcy G.W., 'A planet orbiting 47 Ursae Majoris', *Astrophysical Journal Letters*, **464**, L153, 1996.

Butler P., Marcy G., Fischer D.A., Brown T.M., Contos A.R., Korzennik S.G., Nisenson P. and Noyes R.W., 'Evidence for a multiple companion to Ups Andromedae', *Astrophysical Journal*, **526**, 916, 1999.

Croswell K., *Planet Quest*, Oxford, Oxford University Press, 1997.

Griffin R., 'A photoelectric radial-velocity spectrometer', *Astrophysical Journal*, **148**, 465, 1967.

Lin D.N.C., Bodenheimer P. and Richardson D.C., 'Orbital migration of the planetary companion of 51 Pegasi to its present location', *Nature*, **380**, 606, 1996.

Mayor M. and Queloz D., 'A Jupiter-mass companion to a solar-type star', *Nature*, **378**, 355, 1995.

Naef D., Mayor M., Pepe F., Queloz D., Santos N.C., Udry S. and Burnet M., 'The CORALIE survey for southern extrasolar planets. V. 3 new extrasolar planets, *Astronomy & Astrophysics*, **375**, 205, 2001.

Perryman M.A.C., 'Extrasolar planets: a review article', *Reports in Progress in Physics*, May 2000.

'Les nouvelles planètes', *La Recherche*, 290, September 1996. Dossier spécial.

Future techniques

Angel J.R.P. and Woolf N.J., 'À la recherche de la vie', *Pour la Science*, 224, June 1996.

Burrows A., Saumon D., Guillot T., Hubbard W.B. and Lunine J.I., 'Prospects for detection of extrasolar giant planets by next-generation telescopes', *Nature*, **375**, 299, 1995.

Charbonneau D., Brown T.M., Latham D.W. and Mayor M., 'Detection of planetary transits across a sun-like star', *Astrophysical Journal*, **529**, L45, 2000.

Henry G.W., Marcy G.W., Butler R.P. and Vogt S.S., 'A transiting 51 Peg-like planet', *Astrophysical Journal*, **529**, L41, 2000.

Labeyrie A., 'Des miroirs en pointillés', *La Recherche, 329*, March 2000.

Léger A., Mariotti J.M., Mennesson B., Ollivier M., Puget J.L., Rouan D. and Schneider J., 'Could we search for primitive life on extrasolar planets in our near future? The Darwin project', *Icarus*, **123**, 249, 1996.

Some extraterrestrial reading matter

Cocconi G. and Morrison P., 'Searching for interstellar communications', *Nature,* **184**, 844, 1959.

Drake F.D., 'Project OZMA', *Physics Today*, **14**, 40, 1961.

Frankel C., *La Vie sur Mars*, Paris, Éditions du Seuil, coll. 'Science ouverte', 1999.

Pappalardo R., Head J. and Greeley R., 'L'océan caché d'Europe', *Pour la Science*, 268, February 2000.

Raulin F., Raulin-Cerceau F. and Schneider J., *La Bioastronomie*, Paris, PUF, coll. 'Que Sais-Je?', 1997.

Some internet sites[1]

http://astro.estec.esa.nl/GAIA/

http://astro.estec.esa.nl/IRSI/

http://athene.as.arizona.edu:8000/caao/caao/planets.html

http://cfa-www.harvard.edu/afoe/espd.html

http://exoplanets.org

http://obswww.unige.ch/~udry/planet/planet.html

http://origins.jpl.nasa.gov/

http://www.astrobiology.com/

http://www.estec.esa.nl/spdwww/h2000/html/index.html

http://www.obspm.fr/encycl/encycl.html

[1] The publisher has used its best endeavours to ensure that the URLs for external websites referred to in this book are correct and active at the time of going to press. However, the publisher has no responsibility for the websites and can make no guarantee that a site will remain live or that the content is or will remain appropriate.